T0142797

Intelligent Systems Reference Library

Volume 132

About this Series

The aim of this series is to publish a Reference Library, including novel advances and developments in all aspects of Intelligent Systems in an easily accessible and well structured form. The series includes reference works, handbooks, compendia, textbooks, well-structured monographs, dictionaries, and encyclopedias. It contains well integrated knowledge and current information in the field of Intelligent Systems. The series covers the theory, applications, and design methods of Intelligent Systems. Virtually all disciplines such as engineering, computer science, avionics, business, e-commerce, environment, healthcare, physics and life science are included.

More information about this series at http://www.springer.com/series/8578

Angelo Costa · Vicente Julian
Paulo Novais
Editors

Personal Assistants: Emerging Computational Technologies

Editors
Angelo Costa
Departamento de Informática/Centro ALGORITMI
Escola de Engenharia, Universidade do Minho
Braga
Portugal

Paulo Novais
Departamento de Informática/Centro ALGORITMI
Escola de Engenharia, Universidade do Minho
Braga
Portugal

Vicente Julian
Department of Computer Systems and Computation
Polytechnic University of Valencia
Valencia
Spain

ISSN 1868-4394 ISSN 1868-4408 (electronic)
Intelligent Systems Reference Library
ISBN 978-3-319-87334-3 ISBN 978-3-319-62530-0 (eBook)
DOI 10.1007/978-3-319-62530-0

Softcover reprint of the hardcover 1st edition 2017

Printed on acid-free paper

This Springer imprint is published by Springer Nature
The registered company is Springer International Publishing AG
The registered company address is: Gewerbestrasse 11, 6330 Cham, Switzerland

Preface

It has been shown that the quality of life for people remaining in their own homes is generally better than for those who are institutionalised. Moreover, the cost for institutional care can be much higher than the cost of care for a patient at home. To balance this situation, efforts must be made to move the services and care available in institutions to the home environment. Thus, society poses new challenges, demanding systems that overcome this issue.

Personal assistants (PAs) are a relatively new concept, advancing the Cognitive Orthotics concept, which is only focused on direct assistance to people with cognitive or physical disabilities. The aim is to expand the Cognitive Orthotics area to include complex platforms that include sensors, actuators, monitoring abilities and decision processes.

PA domain contains technologies such as cognitive assistants, multi-agent systems, robotics and applications (such as e-health and e-learning), among others. Essentially, PA is focused on people and their disabilities, providing tools that best fit them using personalisation methods. They have been typically developed to perceive the intrinsic mechanisms of human cognition such as reasoning, learning, memorising, acting and adapting; to discover the thought process leading to each decision; and to build systems that can emulate those thought processes and make decisions or suggestions.

PA can range from a medication reminder to a messaging system that connects its users with their relatives. New developments like the Internet of Things and the increasing amount of computing power that hand-held devices have allowed the development of environments that were until now unavailable through embedded systems. Therefore, there are a lot of implementation options open for development on this area. This book is intended to provide an overview of the research being carried out in the interdisciplinary area of personal assistants and cognitively inspired systems.

The contents of the book were divided into the following parts: Introduction, Reasoning, Health, Personalisation, Robotics, Ethic and Social Issues.

The Introduction presents an overview of the area and the projects that constitute it. The Reasoning presents the knowledge processes that affect PA. The Health presents application of PA in health environments. The Personalisation presents solutions that are directly related to how the PA can adjust to the users. The Robotics presents application of PA through robotic systems. And finally, the Ethic and Social Issues present the legal perspective of how the PA affects the society.

This book counts with international contributions, from countries such as Argentina, Republic of Colombia, Russia, Spain, Portugal, the USA, which provide different perspectives elated to their own culture, being composed by 12 chapters.

In closing we would like to thank the reviewers who helped to increase the excellency of this book.

Braga, Portugal Angelo Costa
Valencia, Spain Vicente Julian
Braga, Portugal Paulo Novais
May 2017

Contents

Part I Introduction

1 A Survey of Cognitive Assistants 3
Angelo Costa, Paulo Novais and Vicente Julian
1.1 Introduction 3
1.2 Cognitive Assistants 5
1.2.1 DayGuide 5
1.2.2 Active@Home 6
1.2.3 CoME 6
1.2.4 DALIA 7
1.2.5 EDLAH2 7
1.2.6 iGenda 9
1.2.7 M3W 9
1.2.8 MyGuardian 10
1.2.9 PersonAAL 12
1.3 Conclusion 13
References 14

Part II Reasoning

2 Argumentation-Based Personal Assistants for Ambient Assisted Living 19
Stella Heras, Javier Palanca and Carlos Iván Chesñevar
2.1 Introduction 19
2.2 Decision-Making and Recommendation 22
2.3 Computational Persuasion 27
2.4 Conclusion and Open Issues 30
References 34

3 **Kidney Care—A Personal Assistant Assessment** 37
Bia Martins, Joao Rei, Miguel Braga, Antonio Abelha,
Henrique Vicente, Joao Neves and Jose Neves
3.1 Introduction 38
3.2 Knowledge Representation and Reasoning 39
3.2.1 Quantitative Knowledge 40
3.2.2 Qualitative Knowledge 44
3.3 A Case Based Approach to Problem Solving 45
3.4 System's Architecture 46
3.5 Case Study 47
3.5.1 Data Processing 48
3.5.2 The CBR Approach to Computing 49
3.6 Conclusion 52
References 53

Part III Health

4 **Visual Working Memory Training of the Elderly in VIRTRAEL Personalized Assistant** 57
Miguel J. Hornos, Sandra Rute-Pérez, Carlos Rodríguez-Domínguez,
María Luisa Rodríguez-Almendros, María José Rodríguez-Fórtiz
and Alfonso Caracuel
4.1 Introduction 57
4.2 Related Work 60
4.3 VIRTRAEL Description 61
4.4 Classification and Memorization of Images Exercise 68
4.4.1 Pilot Study 72
4.4.2 Personalization of the Exercise 72
4.5 Conclusions and Future Work 74
References 75

5 **Personal Robot Assistants for Elderly Care: An Overview** 77
Ester Martinez-Martin and Angel P. del Pobil
5.1 Introduction 77
5.2 Assistive Social Robots 79
5.2.1 Companion Robots 79
5.2.2 Service Robots 84
5.3 Conclusions 88
References 88

Part IV Personalization

6 Personalized Visual Recognition via Wearables: A First Step Toward Personal Perception Enhancement 95
Hosub Lee, Cameron Upright, Steven Eliuk and Alfred Kobsa
6.1 Introduction 96
6.2 Related Work 97
6.3 Personalized Visual Recognition System via Google Glass 99
6.3.1 System Architecture 99
6.3.2 Client 100
6.3.3 Server 100
6.3.4 Workflow 100
6.3.5 Training 100
6.4 Classification 102
6.5 Experiment 1: Person Identification 102
6.5.1 Overview 102
6.5.2 Training Data 103
6.5.3 Finetuning for 20-Class Person Identification 103
6.5.4 Chained Finetuning for 5-Class Person Identification 105
6.5.5 Comparison Between Finetuning and Chained Finetuning 105
6.6 Experiment 2: Object Recognition 107
6.6.1 Overview 107
6.6.2 Training and Validation Data 107
6.6.3 Chained Finetuning for 10-Class Object Recognition 108
6.7 Discussion and Future Work 109
6.8 Conclusion 110
References 111

7 Intelligent Personal Assistant for Educational Material Recommendation Based on CBR 113
Néstor Darío Duque Méndez, Paula Andrea Rodríguez Marín and Demetrio Arturo Ovalle Carranza
7.1 Introduction 114
7.2 Preliminaries 115
7.2.1 Learning Objects (LO), Learning Objects Repositories and Repository Federation 116
7.2.2 Case-Based Reasoning (CBR) 116
7.2.3 Recommender Systems 118
7.2.4 Student Profile 119
7.3 Related Works 119
7.4 Proposed Model 121

7.4.1 CBR Stages in Intelligent Personal Assistant to Recommend Educational Resources . . . 122
7.5 Experiments and Results. . . 124
7.5.1 Study Case . . . 127
7.6 Conclusions . . . 129
References. . . 130

Part V Robotics

8 Characterize a Human-Robot Interaction: Robot Personal Assistance. . . 135
Dalila Durães, Javier Bajo and Paulo Novais
8.1 Introduction . . . 135
8.2 Theoretical Foundations . . . 136
8.2.1 Social Robots . . . 138
8.2.2 Personal Assistance . . . 139
8.3 The Proposed Design . . . 141
8.3.1 Dynamic HRI Monitoring Architecture . . . 143
8.4 Discussions and Conclusions . . . 145
References. . . 146

9 Collaboration Between a Physical Robot and a Virtual Human Through a Unified Platform for Personal Assistance to Humans . . . 149
S.M. Mizanoor Rahman
9.1 Introduction . . . 150
9.2 Related Works . . . 152
9.3 Development of the Personal Assistant Robot and the Virtual Human . . . 153
9.3.1 Development of the Humanoid Robot . . . 154
9.3.2 Development of the Virtual Human. . . 154
9.4 The Unified Platform to Integrate the Operations of the Robot and the Virtual Human . . . 155
9.5 Home-Based Settings to Assist Disabled Persons in Daily Living by the Robot, the Virtual Human and Their Collaboration . . . 156
9.5.1 The Intelligent Robot Assists the Human . . . 156
9.5.2 The Intelligent Virtual Human Assists the Human . . . 158
9.5.3 Collaboration Between the Robot and the Virtual Human to Assist the Human . . . 158
9.5.4 Strategy of Determining the Master and the Follower Agent . . . 161

9.6 Modeling and Measurement of Human Trust in Robot and Virtual Human and Bilateral Trust Between Robot and Virtual Human 162
9.6.1 Trust Modeling 162
9.6.2 Trust Measurement 163
9.7 Evaluation Scheme to Evaluate the Assistance of the Robot, Virtual Human and Their Collaboration to the Disabled Human 165
9.8 Experimental Evaluation of the Quality of the Assistance of the Robot, the Virtual Human and Their Collaboration to the Disabled Human 166
9.8.1 Recruitment of Subjects 166
9.8.2 Experimental Objectives 166
9.8.3 Hypotheses 167
9.8.4 Experimental Procedures 167
9.8.5 Experimental Results 168
9.9 Limitations of the Methods and the Results 173
9.10 Conclusions and Future Works 174
References 175

10 Emotion Detection and Regulation from Personal Assistant Robot in Smart Environment 179
José Carlos Castillo, Álvaro Castro-González, Fernándo Alonso-Martín, Antonio Fernández-Caballero and Miguel Ángel Salichs
10.1 Introduction 179
10.2 The Personal Assistant Robot 180
10.2.1 A Mobile Social Robot 181
10.2.2 A Social Robot with Ears and Eyes 182
10.2.3 A Social Robot with Expressive Capabilities 182
10.3 The Multi-modal Emotion Detection Module 183
10.3.1 Emotion Detection Through Voice Analysis 185
10.3.2 Emotion Detection Through Video Analysis 187
10.3.3 Integration of GEVA and GEFA 188
10.4 The Emotion Regulation Module 189
10.4.1 Musical Emotion Regulation 189
10.4.2 Colour/Light-Based Emotion Regulation 191
10.5 Conclusions 192
References 192

Part VI Ethic and Social Issues

11 EDI for Consumers, Personal Assistants and Ambient Intelligence—The Right to Be Forgotten 199
Francisco Pacheco de Andrade, Teresa Coelho Moreira, Mikhail Bundin and Aleksei Martynov
11.1 Introduction 199
11.2 Electronic Data Interchange 200
11.3 Personal Assistants and Ambient Intelligence 201
11.4 Privacy and Data Protection 202
11.5 Final Remarks 205
References 206

12 Personal Assistants: Civil Liability and Dispute Resolution 209
Marco Carvalho Gonçalves
12.1 Introduction 209
12.2 Protection of Personality Rights 211
12.3 Civil Liability 211
12.3.1 Introduction 211
12.3.2 Types of Illegal Conduct 212
12.3.3 Appreciation of Fault and Damages 213
12.4 Dispute Resolution 215
12.5 Conclusion 218
References 218

Part I
Introduction

Chapter 1
A Survey of Cognitive Assistants

Angelo Costa, Paulo Novais and Vicente Julian

Abstract Cognitive Assistants is a subset area of Personal Assistants focused on ubiquitous and pervasive platforms and services. They are aimed at elderly people's needs, habits, and emotions by being dynamic, adaptive, sensitive, and responsive. These advances make cognitive assistants a true candidate of being used in real scenarios and help elderly people at home and outside environments. This survey will discuss the cognitive assistants' emergence in order to provide a list of new projects being developed on this area. We summarize and enumerate the state-of-the-art projects. Moreover, we discuss how technology support the elderly affected by physical or mental disabilities or chronic diseases.

1.1 Introduction

The term Personal Assistants (PA) is originated from the Ambient Assisted Living (AAL) area that encompasses the advances in the ICT area that are focused in providing direct care on activities of daily living and related tasks. The AAL area focuses in technologies that provide healthcare, assistance and rehabilitation to elderly or disabled people (with cognitive and physical impairments), promoting independent living, active aging and aging in place. Therefore, the need for the distinction is required due to the fact that not all PA technologies belong to AAL and vice-versa. Recently PA have gained traction and there are several projects with interesting results.

A. Costa (✉) · P. Novais
Departamento de Informática/Centro ALGORITMI, Escola de Engenharia, Universidade do Minho, Braga, Portugal
e-mail: acosta@di.uminho.pt

P. Novais
e-mail: pjon@di.uminho.pt

V. Julian
Departamento de Sistemas Informáticos y Computación, Universitat Politècnica de València, Valencia, Spain
e-mail: vinglada@dsic.upv.es

A. Costa et al. (eds.), *Personal Assistants: Emerging Computational Technologies*, Intelligent Systems Reference Library 132, DOI 10.1007/978-3-319-62530-0_1

The need for projects on these areas comes from the increasing numbers of elderly people that need technological solutions on their daily lives [38]. Most of the elderly are fragile and need assistance to perform certain tasks. While most try to overcome these issues however they can and at great physical and psychological cost, some are unable to perform those tasks at all. Moreover, population aged 65 or older is projected to increase from an estimated 520 million in 2010 to nearly 1.5 billion in 2050 [38]. With the growth of elderly population comes the increase of mental and physical diseases like Alzheimer, dementia and mobility problems.

Population Reference Bureau [27] states that about 20 million North Americans assist elderly people performing daily activities, and that 70% need constant care. Monetarily speaking, in 2010 the spending in care of people with dementia ascended to 604 billion dollars, and with the increase of population its expected the increase of costs [38]. The health-care costs dedicated to dementia patients is estimated to be in the UK roughly £26 billion a year [26]. Two-thirds (£17.4 billion) is paid by people with dementia and their families, either in unpaid care or in private social care. Thus it is expected that the continuous increase of the elderly population will produce an economical disadvantage to their families and society.

One way to address this issues is with the use of technological devices that are able to help and simplify the execution of daily tasks. Low cost health-care systems can be built to diminish the burden of the caregivers with minimal transitional periods and with a high level of usability. Technological devices can help to break social and physical barriers that elderly people and caregivers have and enable independent living with a high rate of success.

Governments have already observed that these advances can be very advantageous. The European Union has and keeps promoting and funding projects on this area [14] through several calls to action.

Also, there is the European Innovation Partnership on Active and Healthy Aging (EIP AHA) that is constituted by several actors that work in conjunction to propose new ideas and pursuit technological developments directed to the elderly population; it aims to continually increase the levels of the elderly health condition by 2020 [15].

In terms of policies and protocols, there are some projects that tackle those issues through compilation of information on existing services and private and public initiatives, and conduction of representative surveys within the elderly population.

The AALIANCE project [33] built a guideline for development and research in the AAL area, as well as some policies that should be implemented at the European level. It is clear that there is a great interest in the PA and AAL area, where state of the art advances are currently being produced.

This work aims to provide an overview of the cognitive assistants inside the PA area, focusing on technologies and approaches for aging population in home environments. More specifically, we will present the concept and usage of cognitive assistants, which are constituted by platforms, services and tools that help the users overcome their cognitive disabilities through discreet and ubiquitous devices.

This chapter is structured as follows: Sect. 1.2 presents the cognitive assistants projects and their setting in the PA area; Sect. 1.3 presents the conclusions.

1.2 Cognitive Assistants

Cognitive Assistants (CA) are well integrated in the PA area and constitute a larger portion of that area. The CA have as social goal the production of tools that help people with cognitive disabilities to perform activities of daily living. Therefore, most of the advances produced on this area are software platforms, as the target users do not have any motor disability or only have mild motor disabilities caused by their psychological impairment.

Projects on this area seem to move towards a unified system that is able to interconnect to external services and create an extended technological environment [20]. This environment will be greater that the sum of the parts due to the possibility of data and sensor fusion, thus making available complex information that otherwise was unavailable. As a toy example, and for easy envisioning, we can take two sensors that when not unified could bring a lot of problems that are the smoke and flood detection; if there is a fire the sprinklers will be activated, the flood sensor will be activated and stop the sprinklers thus allowing the fire to spread; this process would be in cycle until one of the sensors stopped working. Therefore, with this example we can observe that interconnected systems can build interesting information when working together.

The following projects are a display of what is being developed currently conceptually and architecturally. They are a small sample that serve the goal of presenting a heterogeneity representation of solutions within the CA area.

1.2.1 DayGuide

The DayGuide project [3] aims to provide reminders associated to locations and guidance, a social platform for share and organization of tasks, through a mobile phone.

It is designed to be used by elderly people suffering from mild cognitive impairment (MCI) in an aging at home perspective. This project is fairly new and shows signs of being on an initial development phase, as showed by the interface on Fig. 1.1.

In terms of operation, the authors present the following toy example: "*When the person with dementia opens the entrance door, s/he receives specific reminders depending on time of the day, outside weather conditions and diary*". The reminders

Fig. 1.1 DayGuide expected interface

are expected to be presented visually and well as aurally. Furthermore, their partnership has the ability to tap to an care-center environment with 60 persons that can be used for tests and validation.

1.2.2 Active@Home

The Active@Home project [1], which is very recent and currently does not present large developments, focuses on active people and how to maintain them active. Their approach is to promote game-based exercises designed for elderly people that engages them through dance and Tai Chi activities. The main goal is to avoid falls by keeping the elderly physically and cognitively exercised, improving their balance.

The technological implementation uses televisions (or large display devices) to show animated virtual characters performing the exercises, and capture the exercise execution by the viewers through wearable sensors. The information of the sensors will help to determine if there is an decay of the physical condition and if there is any undetected health problem. In terms of cognitive assistance, this project helps to detect cognitive illnesses and uses physical exercises to juggle their memory and remember previously done activities.

1.2.3 CoME

The CoME project [2], is a platform for monitoring and interacting with elderly people. The platform counts with wearable sensors to constantly monitor the elderly people and smartphones to visually interact with them and to receive self-reports from them. Moreover, the smartphone will be used to show tutorials about how to perform certain activities and be used by the caretakers to localize and receive health reports from their care-receivers.

In Fig. 1.2 is showed the architecture of the project. We can observe a complex CA platform that includes all the actors and several services that are aimed at maintaining and improving the cognitive status of the system actors. The platform is intended to be deployed at each elderly home, connecting them to a central service that provides information to the formal and informal caretakers, thus promoting the aging in place concept. Therefore, the platform also pushes the idea that it helps the caretakers by relieving them of the burden and stress of the constant supervision.

In terms of technological solutions, Fig. 1.2 shows that it is a typical client-server structure with a high volume of communications between them. It is not specified if there will be any local processing system for the wearables data or the smartphone data.

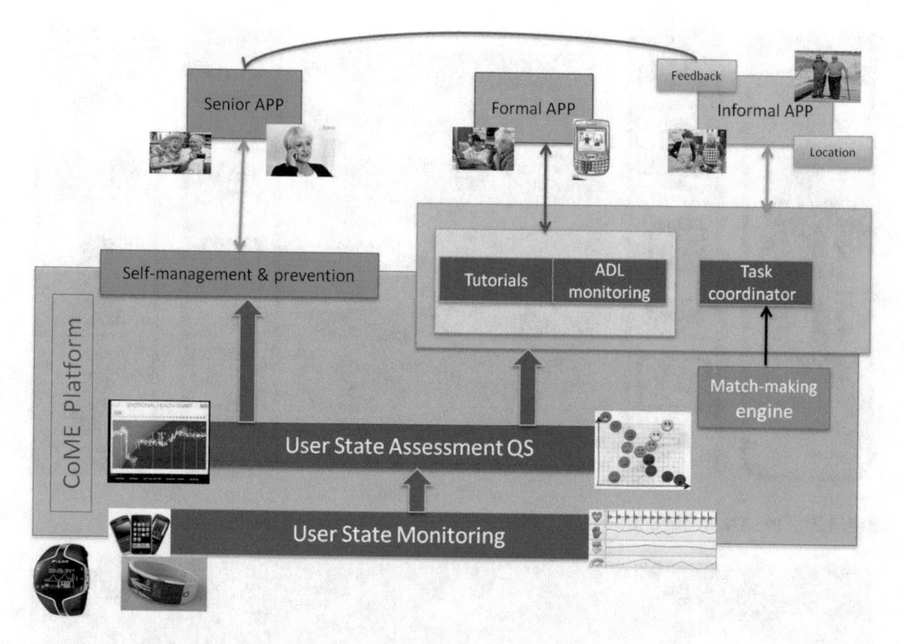

Fig. 1.2 CoME architecture

1.2.4 DALIA

The DALIA project [9, 39] is an wrapper for the ANNE virtual assistant, providing ANNE with the abilities of linear integration with external services and a service-oriented architecture. ANNE is a virtual assistant in the shape of a human-looking avatar endowed with speech and face recognition and speech capabilities.

The combined platforms result in an platform-independent virtual assistant, showed in Fig. 1.3, that has the following abilities:

- Easy communication with family and friends through the virtual assistant;
- Appointments calendar and automatic reminders;
- Documenting events and keeping memories;
- Emergency call and fall detection;
- Localisation of lost items;
- Health status monitoring and motivation for physical or mental activity tasks.

1.2.5 EDLAH2

The EDLAH2 project [4] aims to use gamification procedures on elderly people resorting to tablet technology. By showing appellative visual interfaces, as showed in

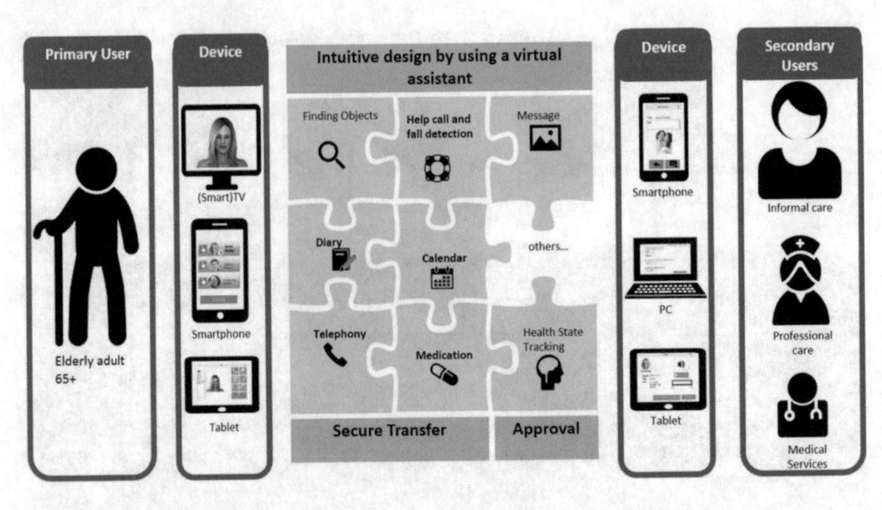

Fig. 1.3 DALIA architecture

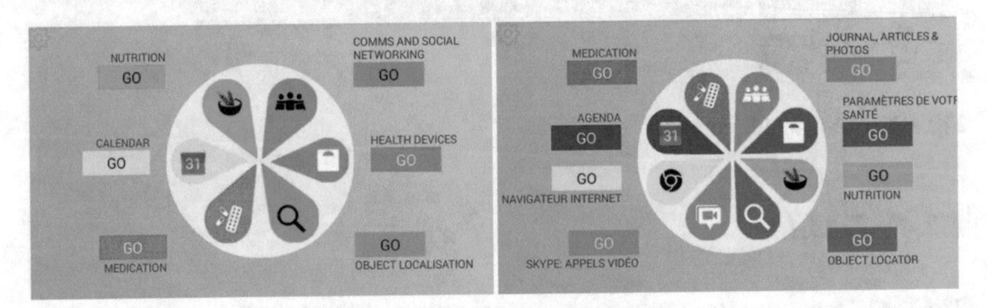

Fig. 1.4 EDLAH2 visual interfaces

Fig. 1.4, it helps elderly people to have a smooth interaction with normally difficult technology.

Being the evolution of the EDLAH project [7], it uses gamification to tap into the basic desires and needs of the user impulses, promoting status and achievements.

The objective is to be as appealing as possible and to have a greater incentive to be used, engaging elderly people into the use of new technologies, as seen in Fig. 1.4. Moreover, it may improve the mental health, physical health, social integration, and self-esteem. It will be employed positive reinforcement suggestions and messages, that cater to the user progress and imply further usage of the applications and tablet.

In terms of gamification the execution process will give awards gained by using other services, e.g. web browsing, games, visualizing photos, and those points will be put in a leader board constituted by a closed set of friends, or the general community, or just personal classification (if one wishes not to be compared).

This PA may be considered as active, as it requires direct interaction from the users for it to work and influence them. Nacke et al. [22] refer that casual and continuous gaming among elderly people using digital devices show a noticeable increase in cognitive development and physical development. Thus this project is an assistant in terms of assisting the cognition health of its users.

1.2.6 iGenda

The iGenda project [10–13] is a cognitive assistant that focuses on elderly people and their caregivers through management of daily events and activities. The idea behind this project is to use technology to tackle the issues of forgetting activities and by remembering them to the elderly, they shuffle through their memories, thus exercising their cognition. Furthermore, the iGenda promotes playful activities (physical and psychological) that enable the elderly to have an active aging.

The iGenda works in two fronts: active aging and aging at home. By taking into account the users' profile and their health issues, the iGenda plans the appropriate activities to each user, being them more outgoing or more homely.

The idea behind using events and activities is supported by several studies that defend using activities and social interaction help to maintain the cognition levels and help stopping the Alzheimer's advance [5, 6, 16, 19, 25, 28–30, 34–37].

The iGenda also is designed to the caregivers, allowing them to manage the activities, schedule new activities or visualise the elderly vital signs (when a wearable device is used). Therefore, it decreases the caregivers' stress coming from the constant care-receiver monitoring and allowing them to monitor several people at the same time.

In terms of operation, the iGenda runs on smartphones or devices with web browsers and is transparent to the users, only warning them when there is an critical interaction required or if the system detects that the users are not performing the activity that was planned at that time.

The iGenda is implemented over multi-agent platforms that are modular and easily scalable. An overview of the architecture is showed in Fig. 1.5.

Currently the iGenda developments are in the emotion detection area and the persuasive area. The goal is to gather unbiased information from the users emotional state and how they respond to certain activities suggestions, and to justify using natural language and human concepts the reason why one activity was suggested over others through persuasion processes.

1.2.7 M3W

The M3W project [23, 24] aim is to develop a mental wellness tool for self-usage to measure and visualise mental changes and tendencies, and to give indications, alarms or reports This tool is directed to the elderly and their relatives, friends, and physicians.

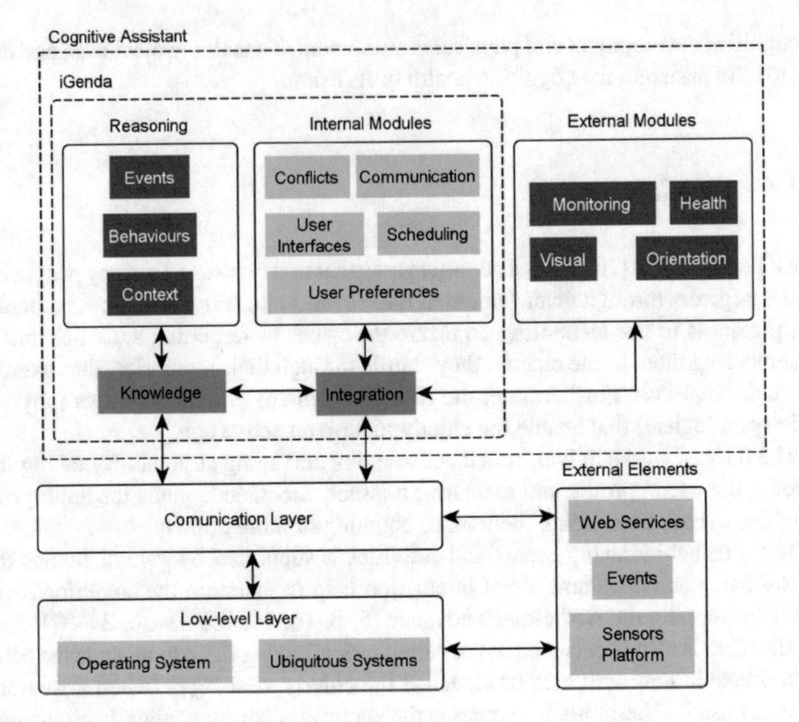

Fig. 1.5 iGenda architecture

The goal is to create an personal health record through historic values and show the positive or negative progression of the health condition.

To evaluate the users, this project uses cognitive games that are designed directly to improve the cognition of the users following the works of [8, 21, 31]. They improve the attention, executive functions (decision making, mental flexibility, planning, and problem solving), memory (visual memory, spatial memory, and working memory), and language. The architecture of execution can be seen in Fig. 1.6.

Like the EDLAH2 project, this CA requires direct interaction from the users, being in the paradigm of assisting through interaction.

1.2.8 MyGuardian

The MyGuardian project [18, 32] aims to use technology to facilitate the elderly mobility, keeping their autonomy and dignity. The generated tool helps the users and caregivers by guiding the elderly and reporting situations of confusion or risk to the caretakers. The caretakers have the additional feature of coordinating the daily tasks step-by-step. To this, the interfaces are simple and easy to understand, as seen in Fig. 1.7.

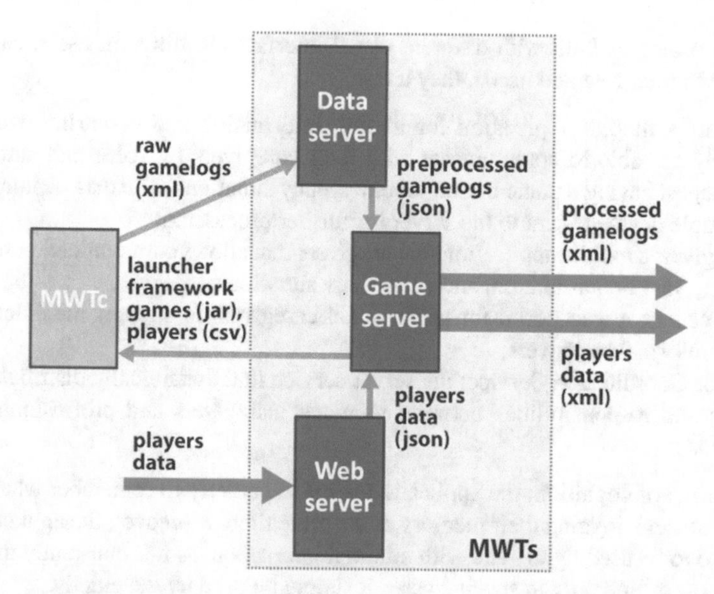

Fig. 1.6 M3W architecture

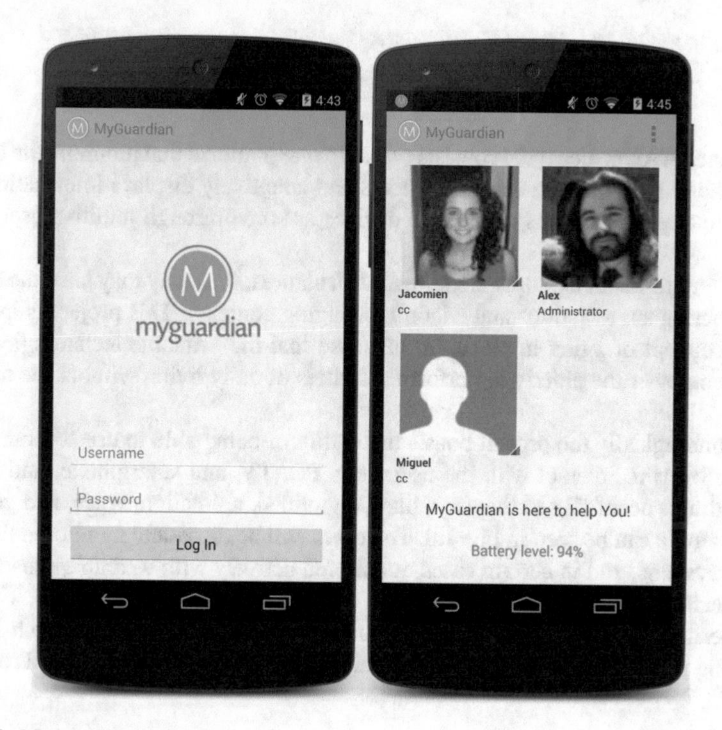

Fig. 1.7 MyGuardian mobile interfaces

This project is built with different visual interfaces to different users, each one specific to their role and needs, they are:

- Senior: a mobile application for trading information and capturing data. The elderly are able to communicate with their caregivers by voice and automated messages, having a panic button for emergency situations. The data captured by a wearable device is sent to the server for further processing;
- Caregiver: a mobile application for caregivers that allows communication with the elderly, tasks coordination, and emergency situations;
- Call-center: a web portal for professional caregivers to support the elderly and their informal caregivers;
- Senior Coordination Service: the server service that dynamically distributes care tasks and responsibilities between volunteer caregivers and professional caregivers.

In terms of cognition, the application helps the elderly to remember where they should be next, jogging their memory at the same time. Moreover, the application is intended to be used lightly and with minimal interaction, as it is constantly monitoring the users, and only in specific cases it should be used by the elderly.

1.2.9 PersonAAL

The PersonAAL project [17] consists of a software platform that monitors the behavior of the elderly through sensors systems and adaptively displays information and health-related suggestions on various devices at home through intuitive user interfaces.

The environment explores model-based structures, thus they may have the ability of predicting some actions and adapt to changing contexts. This project is inserted in the concept of aging in place, in the sense that the available technological elements empower the elderly to perform activities of daily living without the need of additional help.

Technologically the project boasts the ability of being able to use several different platforms to interact with the users, e.g. PC, TV, and smartphone, and different mediums according to the users likes, e.g. video, animation, voice, and gesture. One example can be seen in Fig. 1.8. The users will be discretely monitored through sensor systems, and in certain cases, monitored actively with wearable sensors that specifically monitor health points.

In terms of usage, although the interaction with the user is very much active, requiring response from the users, it is mainly transparent to the users until required by them.

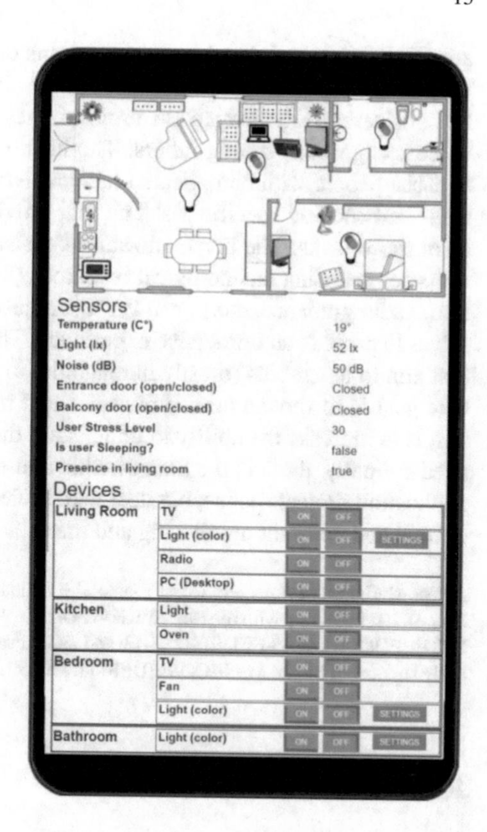

Fig. 1.8 PersonAAL example application

1.3 Conclusion

This work presents the state of the art of the CA area through some very recent projects, here the oldest one is 6 years old. The CA showed touch an array of different social areas and show that there is still a lot of issues to resolve. The complexity of the human being way of think and act is very complex and there is still no way to translate them to computer systems, and the current developments of these translations efforts resolve into long term and very complex projects. From this work we may observe that only to emulate basic aspects of activities scheduling and daily tasks execution takes a lot of effort.

We believe that we are witnessing a new dawn in terms of personal and cognitive assistants, with ever increasing projects and tools. These are areas that are now being extended due to the society high demand of products directed to the elderly. What we foresee is the adoption of these tools by young adults as they become more heterogeneous and features diverse.

The Personal Assistants area is ample and several projects fit on it due to the scope and versatility of the theme. The application area is mostly composed by reasoning, health, personalization and robotics, ethic and social issues projects. The reasoning area deals with high-level implementations of PA. In this area are projects that are very domain specific but essential to the development of PA. Although they may not be considered fully fledged PA projects, they can contribute to other projects, like voice recognition, among others. The health domain consists in projects that aim to assist people with complicated medical issues that are unable to perform certain tasks and need an specific assistant that provides help in directly or indirectly perform those tasks. The personalization consists in projects that aim to enhance the users senses using technological resources (e.g. augmented reality, active information, audio guidance, etc.) or in PA that adjust to the users' profile and adapt themselves to provide an immersive experience. The robotics domain consists in projects that aim to use robots (mostly humanoids) to provide direct assistance to the users. One goal is to show a human presence and interact physically with users. Another aim is to increase the ability to understand the environment and the users and their needs. Finally, there is the area of ethic and social issues that does not present any application projects per se, but deal with societal issues like the ethical ramifications of total and constant monitoring and the society position over sensible topics.

Acknowledgements Angelo Costa thanks the Fundação para a Ciência e a Tecnologia (FCT) the Post-Doc scholarship with the Ref. SFRH/BPD/102696/2014. This work is also supported by COMPETE: POCI-01-0145-FEDER-007043 and FCT Project Scope: UID/CEC/00319/2013 and partially supported by the MINECO/FEDER TIN2015-65515-C4-1-R.

References

1. Active@home. http://www.active-at-home.com (2017). Accessed on 07 April 2017
2. Come. http://come-aal.eu (2017). Accessed on 07 April 2017
3. Dayguide. https://www.dayguide.eu (2017). Accessed on 07 April 2017
4. Edlah2. http://www.edlah2.eu (2017). Accessed on 07 April 2017
5. Abbott, R.D., White, L.R., Ross, G.W., Masaki, K.H., Curb, J.D., Petrovitch, H.: Walking and dementia in physically capable elderly men. Jama **292**(12), 1447–1453 (2004)
6. Akbaraly, T., Portet, F., Fustinoni, S., Dartigues, J.F., Artero, S., Rouaud, O., Touchon, J., Ritchie, K., Berr, C.: Leisure activities and the risk of dementia in the elderly results from the three-city study. Neurology **73**(11), 854–861 (2009)
7. Anagnostopoulos, G.G., Deriaz, M.: Automatic switching between indoor and outdoor position providers. In: 2015 International Conference on Indoor Positioning and Indoor Navigation (IPIN). Institute of Electrical and Electronics Engineers (IEEE) (2015). doi:10.1109/ipin.2015.7346948
8. Anguera, J.A., Boccanfuso, J., Rintoul, J.L., Al-Hashimi, O., Faraji, F., Janowich, J., Kong, E., Larraburo, Y., Rolle, C., Johnston, E., Gazzaley, A.: Video game training enhances cognitive control in older adults. Nature **501**(7465), 97–101 (2013). doi:10.1038/nature12486
9. Auinger, K., Kriegel, J.: Identifikation von nutzeranforderungen durch kreativtechniken am beispiel des europäischen aal projekts dalia. In: e-Health 2014— Informationstechnologien und Telematik im Gesundheitswesen, pp. 277–282 (2013)
10. Carneiro, D., Gomes, M., Costa, Â., Novais, P., Neves, J.: Enriching conflict resolution environments with the provision of context information. Expert Syst. n/a–n/a (2013). doi:10.1111/exsy.12049

11. Costa, Â., Castillo, J.C., Novais, P., Fernández-Caballero, A., Simoes, R.: Sensor-driven agenda for intelligent home care of the elderly. Expert Syst. Appl. **39**(15), 12192–12204 (2012). doi:10.1016/j.eswa.2012.04.058
12. Costa, A., Julián, V., Novais, P.: Advances and trends for the development of ambient-assisted living platforms. Expert Syst. **34**(2), e12, 163 (2016). doi:10.1111/exsy.12163
13. Costa, A., Novais, P., Simoes, R.: A caregiver support platform within the scope of an ambient assisted living ecosystem. Sensors **14**(3), 5654–5676 (2014). doi:10.3390/s140305654
14. EuroHealthNet: European policies and initiatives. http://www.healthyageing.eu/european-policies-and-initiatives (2017). Accessed on 07 April 2017
15. European Commission: European innovation partnership on active and healthy ageing. https://ec.europa.eu/eip/ageing/home_en (2017). Accessed on 07 April 2017
16. Fabrigoule, C., Letenneur, L., Dartigues, J.F., Zarrouk, M., Commenges, D., Barberger-Gateau, P.: Social and leisure activities and risk of dementia: a prospective longitudinal study. J. Am. Geriatr. Soc. **43**(5), 485–490 (1995)
17. Ghiani, G., Manca, M., Paternò, F., Santoro, C.: End-user personalization of context-dependent applications in AAL scenarios. In: Proceedings of the 18th International Conference on Human-Computer Interaction with Mobile Devices and Services Adjunct—MobileHCI'16. ACM Press (2016). doi:10.1145/2957265.2965005
18. Gustarini, M., Wac, K.: Smartphone interactions change for different intimacy contexts. In: Lecture Notes of the Institute for Computer Sciences, Social Informatics and Telecommunications Engineering, pp. 72–89. Springer International Publishing (2014). doi:10.1007/978-3-319-05452-0_6
19. Karp, A., Paillard-Borg, S., Wang, H.X., Silverstein, M., Winblad, B., Fratiglioni, L.: Mental, physical and social components in leisure activities equally contribute to decrease dementia risk. Dement. Geriatr. Cogn. Disord. **21**(2), 65–73 (2006)
20. Kunnappilly, A., Seceleanu, C., Lindén, M.: Do we need an integrated framework for ambient assisted living? In: Ubiquitous Computing and Ambient Intelligence, Lecture Notes in Computer Science, vol. 10070, pp. 52–63. Springer Nature (2016). doi:10.1007/978-3-319-48799-1_7
21. McNab, F., Varrone, A., Farde, L., Jucaite, A., Bystritsky, P., Forssberg, H., Klingberg, T.: Changes in cortical dopamine D1 receptor binding associated with cognitive training. Science **323**(5915), 800–802 (2009). doi:10.1126/science.1166102
22. Nacke, L.E., Nacke, A., Lindley, C.A.: Brain training for silver gamers: effects of age and game form on effectiveness, efficiency, self-assessment, and gameplay experience. CyberPsychol. Behav. **12**(5), 493–499 (2009)
23. Pataki, B., Hanák, P., Csukly, G.: Computer games for older adults beyond entertainment and training: Possible tools for early warnings—concept and proof of concept. In: Proceedings of the 1st International Conference on Information and Communication Technologies for Ageing Well and e-Health. Scitepress (2015). doi:10.5220/0005530402850294
24. Pataki, B., Hanák, P., Csukly, G.: Surpassing entertainment with computer games: online tools for early warnings of mild cognitive impairment. In: Communications in Computer and Information Science, pp. 217–237. Springer Nature (2015). doi:10.1007/978-3-319-27695-3_13
25. Podewils, L.J., Guallar, E., Kuller, L.H., Fried, L.P., Lopez, O.L., Carlson, M., Lyketsos, C.G.: Physical activity, apoe genotype, and dementia risk: findings from the cardiovascular health cognition study. Am. J. Epidemiol. **161**(7), 639–651 (2005)
26. Pool, J.: Alzheimer's Society Guide to the Dementia Care Environment. Alzheimer's Society (2015)
27. Population Reference Bureau: America's aging population. Technical report. http://www.prb.org/pdf11/aging-in-america.pdf (2011)
28. Rovio, S., Kåreholt, I., Helkala, E.L., Viitanen, M., Winblad, B., Tuomilehto, J., Soininen, H., Nissinen, A., Kivipelto, M.: Leisure-time physical activity at midlife and the risk of dementia and alzheimer's disease. Lancet Neurol. **4**(11), 705–711 (2005)
29. Sanders, A., Verghese, J.: Leisure activities and the risk of dementia in the elderly. Res. Pract. Alzheimer's Dis. **12**, 54–58 (2007)

30. Scarmeas, N., Levy, G., Tang, M.X., Manly, J., Stern, Y.: Influence of leisure activity on the incidence of alzheimer's disease. Neurology **57**(12), 2236–2242 (2001)
31. Smith, G.E., Housen, P., Yaffe, K., Ruff, R., Kennison, R.F., Mahncke, H.W., Zelinski, E.M.: A cognitive training program based on principles of brain plasticity: results from the improvement in memory with plasticity-based adaptive cognitive training (IMPACT) study. J. Am. Geriatr. Soc. **57**(4), 594–603 (2009). doi:10.1111/j.1532-5415.2008.02167.x
32. Tsiourti, C., Joly, E., Wings, C., Moussa, M.B., Wac, K.: Virtual assistive companions for older adults: qualitative field study and design implications. In: Proceedings of the 8th International Conference on Pervasive Computing Technologies for Healthcare. ICST (2014). doi:10.4108/icst.pervasivehealth.2014.254943
33. Van Den Broek, G., Cavallo, F., Wehrmann, C.: AALIANCE Ambient Assisted Living Roadmap (Ambient Intelligence and Smart Environments). IOS Press (2010)
34. Verghese, J., LeValley, A., Derby, C., Kuslansky, G., Katz, M., Hall, C., Buschke, H., Lipton, R.B.: Leisure activities and the risk of amnestic mild cognitive impairment in the elderly. Neurology **66**(6), 821–827 (2006)
35. Wang, H.X., Karp, A., Winblad, B., Fratiglioni, L.: Late-life engagement in social and leisure activities is associated with a decreased risk of dementia: a longitudinal study from the kungsholmen project. Am. J. Epidemiol. **155**(12), 1081–1087 (2002)
36. Wang, J., Zhou, D., Li, J., Zhang, M., Deng, J., Tang, M., Gao, C., Lian, Y., Chen, M.: Leisure activity and risk of cognitive impairment: the Chongqing aging study. Neurology **66**(6), 911–913 (2006)
37. Wilson, R.S., De Leon, C.F.M., Barnes, L.L., Schneider, J.A., Bienias, J.L., Evans, D.A., Bennett, D.A.: Participation in cognitively stimulating activities and risk of incident alzheimer disease. Jama **287**(6), 742–748 (2002)
38. World Health Organization: Global health and aging. Technical report, National Institute on Aging, National Institutes of Health, U.S. Department of Health and Human Services (2011)
39. Zach, H., Peinsold, P., Winter, J., Danner, P., Hatzl, J.: Using proxy re-encryption for secure data management in an ambient assisted living application. In: Lecture Notes in Informatics—Proceedings of the Open Identity Summit 2015, pp. 71–83 (2015)

Part II
Reasoning

Chapter 2
Argumentation-Based Personal Assistants for Ambient Assisted Living

Stella Heras, Javier Palanca and Carlos Iván Chesñevar

Abstract Personal assistants may help the elderly population to live independently and improve their welfare in ambient assisted living environments. However, although there are current proposals already developed both in academic and commercial domains, these systems are still far from being established on the daily lives of the general population. Argumentation technologies can help to deal with open challenges in this domain. In this chapter, we explore the connection between the related areas of argumentation, recommendation, decision-making and persuasion, and we review related work that can play an important role on the development of the next generation of personal assistants for ambient assisted living.

2.1 Introduction

Nowadays, many studies claim that *Time* is people's most valuable asset [33]. Consequently, Personal Assistants (PAs), which are able to assist people in many tasks and save them time, have recently gained much success. However, the concept that most people have about a PA has deeply changed over time. While in the 20th century PAs were seen as that secretary or administrative staff that all the businessmen or celebrities had, if today we ask someone if they know any PA, it is very likely that they will answer *Siri*,[1] *Cortana*[2] or *Google Now*.[3] The conception of *time* has also

[1]http://www.apple.com/ios/siri.
[2]https://www.microsoft.com/en-us/windows/cortana.
[3]https://www.google.com/search/about.

S. Heras(✉) · J. Palanca
Universitat Politècnica de Valencia, Camino de Vera s/n, 46022 Valencia, Spain
e-mail: sheras@dsic.upv.es

J. Palanca
e-mail: jpalanca@dsic.upv.es

C.I. Chesñevar
Institute for Research in Computer Science and Engineering (ICIC),
UNS - CONICET, San Andrés 800, Bahía Blanca, Argentina
e-mail: cic@cs.uns.edu.ar

A. Costa et al. (eds.), *Personal Assistants: Emerging Computational Technologies*,
Intelligent Systems Reference Library 132, DOI 10.1007/978-3-319-62530-0_2

undergone an arguably slight change. The enormous technological advances of the last decades have gone hand in hand with a progressive aging of the population, and an unprecedented availability of information. Nowadays, people do not have time to learn how to take advantage of the enormous possibilities that technologies and the big volumes of data available on the Internet may offer them. Furthermore, they not only want to have more time, they want more *quality time*, and this means welfare in terms of health and happiness.

So here is where these new and intelligent software PAs can play their crucial role. They could act on people's behalf (taking care of us or our relatives, helping us in our duties, checking our agenda, etc.), and all this in the most efficient way and with the highest quality. Paradoxically, their level of acceptance by the general population is still very low. Although many have PAs in their mobile devices, for instance, they do not feel them as really necessary. As pointed out in [15], PAs *reductionist* architecture may be the cause for their slow establishment in our daily lives. They are still no more than '*an interface to a collection of discrete and effectively independent underlying functionalities*', but this '*does not add up to intelligence*'. Technologies for personal assistance are still in their infancy and there is a huge room for new approaches and cross-fertilization from other related research areas.

In this chapter, we will focus on the specific domain of PAs for Ambient Assisted Living (AAL). AAL deals with the application of ICT-based solutions for aging well, one of the main pillars of the European Commission roadmap.[4] By 2060, 1 in 3 Europeans will be over 65 and the ratio of working people to the 'inactive' others will be shifting from 4 to 1 today to 2 to 1 [17]. This imminent aging of the population will entail a series of problems that must be addressed from this very moment. Among them, the combination of socialization problems, physical and/or psychological limitations, and the need of specific care raises important challenges for PAs in AAL.

On the one hand, PAs may help the elderly population to live independently and improve their welfare. These technological devices can bring together the elderlies' relatives or friends to make the elderly person feel accompanied. Also, they can manage and track elderlies' activities, show them reminders to take pills, warn caregivers when they are nor feeling well or need help, etc. However, elders did not grown in the current technological age and most have no experience with PAs. Notwithstanding, most show some interest in operating new devices and in learning new things, although they have a general distrust towards computers when they have not been taught to use them [1]. The issue with the current PAs is that they are not truly designed for the elderly, and most will abandon such applications if they are not forced to use them.

To achieve the ultimate goal that the elders actually use PAs in their daily lives, they need to build a relationship of trust between them and the application. Unfortunately, people tend to trust the information that is presented by digital systems even if it is not true or it is incomplete, and when people realize that they were tricked by the system, they stop using it [35]. It is crucial that PAs present truthful infor-

[4]http://www.aal-europe.eu/.

mation, in an understandable and simple way, personalized and appealing for their users. A growing body of work in artificial intelligence has been devoted to deal with research challenges in this domain, and the argumentation research community has also echoed this trend [34]. However, this is still a novel research area with few contributions to date. In this chapter, we review recent advances on computational argumentation in this domain and discuss how these approaches can be successfully applied to overcome research challenges for PAs in AAL settings.

Concretely, the usual operation of a PA in AAL can be seen as a reasoning process with several steps, as depicted in Fig. 2.1. First, the system receives input from different sources of information, like environmental context registered by the PA's sensors (e.g. weather conditions, localization), the user model (e.g. objectives, preferences,

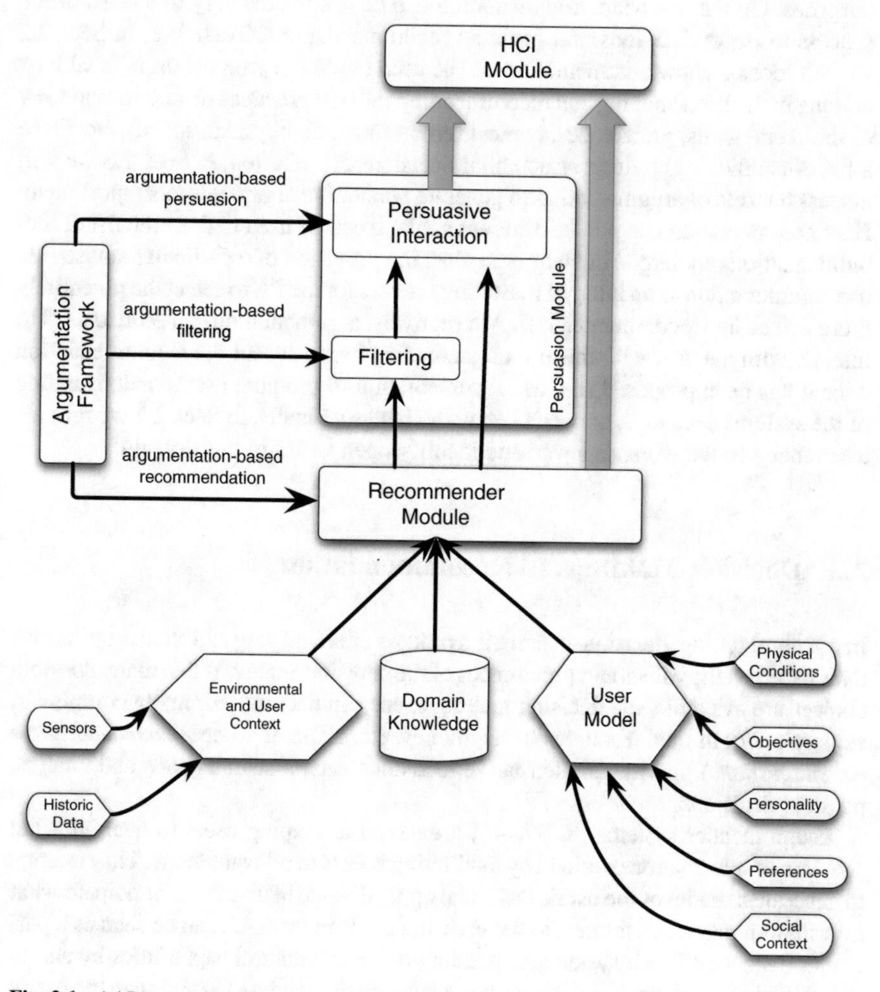

Fig. 2.1 AAL reasoning process

personality, physical conditions, social context), and other possible general knowledge about the domain (e.g. clinical guidelines). Then, it processes these data and generates an advice or recommendation depending on the PA's purpose. Finally, the recommendation is provided to the user. Different styles of human-computer interaction can be followed. For instance, the system may automatically make a decision for the user and directly perform an action on her behalf (e.g. schedule an activity). Other less intrusive approach may be to generate an alert (e.g. a reminder, a warning) that tries to persuade the user to do something, but leaving the final decision of performing it or not at her will. Also, an alternative may also be to allow the user to interact with the system throughout the reasoning process, just to provide more information or to allow her to correct the PA's reasoning (make the PA scrutable).

In this process, argumentation can be applied at different steps and with different purposes. On the one hand, argumentation can be used at the very first steps of the process to make decisions and generate recommendations. Therefore, in Sect. 2.2 we will focus on how argumentation can be used by PAs to promote users' health by helping in decision-making and recommending the best practices or activities in view of the users' goals, preferences, context (environmental, physical, social), etc. Once a list of candidate activities or potential decisions are selected, in Sect. 2.3 we will discuss the role of argumentation to generate persuasive arguments to support them. Here argumentation can pursue different goals. It can be used to filter the list of recommendations and argue in favor or against the provision of a particular advise (i.e. use argumentation as an internal reasoning process for the PA to select the potentially more persuasive recommendation). Alternatively, argumentation can be also used to interact with the user with the aim of persuading her to accept the recommendation once it has been proposed (i.e. use argumentation to promote user's understanding of the systems decisions, or user's behavior change). Finally, in Sect. 2.4 we provide a summary on the work reviewed and identify open issues in this domain.

2.2 Decision-Making and Recommendation

In a general setting, decision-making is a process of identifying and choosing alternatives based on the values and preferences of the decision-maker. When many possible choices are available, the decision making space can become extremely complex to explore under limited resources (time, money, etc.). That is when *recommendations* (or *suggestions*) provide an alternative to reduce such a search space and analyze possible decisions.

Recommender systems [8, 32, 45] are aimed at helping users to deal with the problem of information overload by facilitating access to relevant items. They attempt to generate a model of the user's tasks and apply diverse heuristics to anticipate what information may be of interest to the user. In fact, PAs for AAL can be seen as a particular instance of such systems, expanding the user's natural capabilities by acting as intelligence or memory augmentation mechanisms, tending to minimize the user's cognitive effort (something valuable for human beings in general, and in particular for elderly people).

Recommender systems technologies usually operate by creating a model of the users' preferences or tasks with the purpose of facilitating access to items that the user might find useful. In this respect, many of such systems attempt to anticipate the user's need, providing assistance proactively. In order to come up with recommendations or suggestions, conventional recommender systems rely on similarity measures between users or contents, computed on the basis of methods coming either from the information retrieval or from the machine learning community.

Two main techniques have been used to compute recommendations: *content based* and *collaborative filtering*. Content-based recommenders rely on the premise that user's preferences tend to persist through time, using machine-learning techniques to generate a profile of the user based on ratings she provided in the past. Collaborative filtering recommenders, on the other hand, are based on the assumption that users' preferences are correlated, keeping a pool of users' profiles, so that when decisions are to be made in the context of the active user, she can be presented with items which strongly correlate with other items from similar users in the past.

An important deficiency of recommendation technologies in general is their limited ability to *qualitatively* exploit data, giving rise to a number of issues:

- *Exposing underlying assumptions and providing rationally compelling arguments*: while recommendations may be simple pointers or hints in many situations, it is easy to come up with scenarios in which the user may need further evidence before taking a course of action (e.g. "take a nap after lunch" is a recommendation which is might not be very helpful if not complemented with "since you're over 60 years old and you have high blood presure.").
- *Dealing with the defeasible nature of users' preferences*: modeling the dynamics of user preferences can help to keep a PA for AAL properly up-to-date, without disregarding decisions made by the user in the past.
- *Approaching trust and trustworthiness*: Recommendation technologies are increasingly gaining importance in commercial applications, including PAs. However, most existing systems simply focus on tracking a customers interests and make suggestions for the future without a contextualized justification. As a result the user is unable to evaluate the reasons that led the system to present certain recommendations.

Argumentation [47] provides a natural approach to model the problems discussed before. The research community in this steadily growing discipline in Artificial Intelligence has consolidated itself in the last twenty years, providing interesting contributions in different areas such as multiagent systems [3], social networks [23, 25], machine learning [9], and intelligent decision making [18], among many others.

Given a potentially inconsistent knowledge base, an *argument* is a collection of facts and rules that provide a rational support to reach some conclusion or claim. An argument can be attacked by other *counterarguments*, which can be attacked on their turn. This results in a graph-like structure from which different arguments

Table 2.1 Running example

(1) If you need to take pill *A*, then call pharmacy *X*
(2) If a pharmacy is closed, you cannot get pills there
(3) If you need to take pill *B*, then call pharmacy *Y*
(4) If it is Saturday, pharmacy *X* is closed
(5) Today is Saturday
(6) You need to take pill *A*
(7) You need to take pill *B*

may be considered to prevail (so-called *warranted* arguments), according to different argumentation semantics. Argumentation frameworks may include different additional features such as knowledge capabilities for representing uncertainty, preference orderings among arguments, as well as other notions such as provenance and audience [47].

For the sake of example, for the case of a PA for AAL we could have the over-simplified knowledge base associated with medicines to be taken by some elderly person shown in Table 2.1:

This toy example shows that this PA can obtain an argument for calling pharmacy *Y* (from items 3 and 7), whereas the argument for calling pharmacy *X* (based on items 6 and 1) is defeated by another argument obtained from items 4 and 5. Hence, contacting pharmacy *X* should be aborted.

Traditional expert systems [20] rely on if-then rules in order to draw conclusions, but lack of the ability to perform commonsense reasoning in the presence of contradictory or potentially inconsistent information (as in the case of the knowledge base given in Table 2.1). As pointed out in a seminal work by McCarthy [38], this lack makes them *brittle*. By this is meant that they are difficult to expand beyond the scope originally contemplated by their designers, and they usually do not recognize their own limitations. Argumentation is intended to overcome such limitations, as it is based on *defeasible* rules, which provide tentative reasons to arrive to conclusions (thus, the rule *'If you need to take pill A, then call pharmacy X.'* should be understood as a tentative rule, which may be subject to the presence of exceptions). Argumentation deals with contradictory information by allowing the emergence of different conflicting arguments (as it usually happens in human reasoning processes), and deciding which arguments are ultimately to be accepted based on a particular semantics.

Figure 2.2 summarizes the different aspects of argument-based reasoning [4]. Starting with a knowledge base containing potentially inconsistent information, we are able to infer different, potentially inconsistent arguments $A_1, A_2, \ldots, A_n$ (depicted traditionally as triangles) which account as tentative proofs to reach conclusions q_1, $q_2, \ldots, q_n$, respectively (Fig. 2.2b). Conflicts among arguments may arise; some arguments will defeat other arguments (Fig. 2.2c), based on some preference criterion on arguments. In order to determine which arguments ultimately prevail, there are

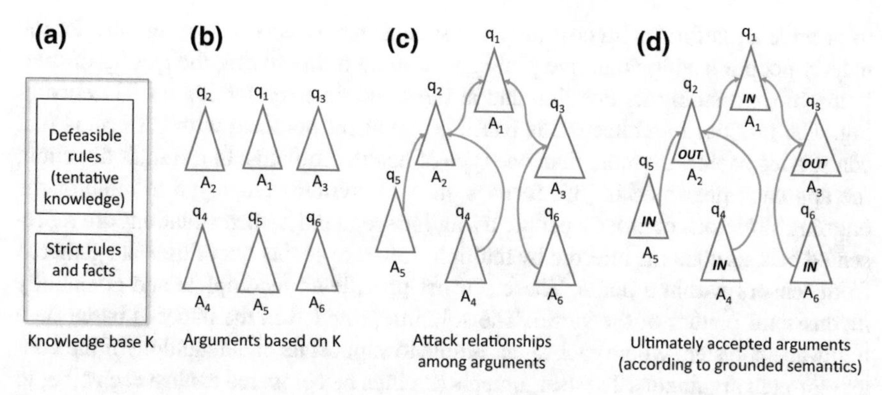

Fig. 2.2 Conceptual view of the different stages involved in the argumentation process

different argumentation semantics which allow to determine which arguments are considered to be *warranted*. For example, according to*grounded skeptical semantics* (Fig. 2.2d) an argument is ultimately accepted (labeled as IN) if it has no arguments defeating it (e.g. arguments A_4, A_5 and A_6), or if all the arguments that attack it are defeated by other arguments which are IN (e.g. as the case for argument A_1). Otherwise, the argument is labeled as OUT (not accepted).

PAs for AAL can benefit from argument-based decision making in many respects. Argumentation theory provides a sound framework for decision making under uncertainty and potentially incomplete information. For the case of PAs, arguments can provide support for different pieces of information, which can be categorized according to their epistimic status (this approach was first introduced in ArgueNet [7], an argument-based web search engine). Argumentation can be also integrated with different formalisms for handling uncertainty (e.g. using possibilistic logic or probabilities), which may also contribute to a richer model for decision making. Furthermore, as explained before, argumentation provides a natural way of tracing the underlying reasoning process carried out to make a decision (instead of adopting a black-box approach, as in many current PAs systems).

Research in argumentation has provided interesting advances which might be helpful for enhancing qualitative reasoning capabilities in PAs for AAL. In [29] an argumentation-based approach to aggregate clinical evidence is proposed. These evidence come from multiple sources, such as randomized clinical trials, systematic reviews, meta-analyses, network analyses, etc., and arguments are used to combine them and decide which treatment outperforms the others or which are equal (can be applied indistinctly). In [46], the authors show how an argumentation-based reasoning service can be used in in a PA for travel services, based on a BDI agent implemented with a Jadex platform. The travel assistant agent illustrates how BDI and argumentation approaches can be effectively integrated in a working system developed with freely available technologies. In [34], the authors discuss engineering aspects for an agent-based AAL system for the home environment using argumentation for decision making. The special requirements of the proposed system are to provide a platform with cost-effective specialized assisted living services for the elderly people having cognitive problems, aiming at improving the quality of their home life, extending its duration and at the same time reinforcing social networking. The proposed architecture is based on an agent platform with PA agents that can service users with more than one type of health problems. In a similar direction, the approach presented in [39] focuses on AAL systems employed to monitor the ongoing situations of elderly people living independently. Such situations are represented here as contexts inferred by multiple software agents out of the data gathered from sensors within a home. These sensors provide an incomplete and potentially inconsistent picture of the world. The solution provided in the paper is based on a multi-agent system where each agent is able to support its understanding of the context through arguments. These arguments can then be compared against each other to determine which agent provides the most reliable interpretation of the reality under observation.

Clearly, an appropriate integration of *context-adaptable criteria* is very important in order to get a PA for AAL working properly. Following the previous example, the recommendations to be provided associated with pills and medicines are subject to different context issues: if the pills are to be taken in the night, some kind of backlight might be needed in the PA device in order to facilitate an easier reading of messages for an elderly person, whereas in daylight the backlight is not required. Similarly, if the person is moving to a different city, the PA's GPS system should provide new context information in order not to look for pharmacies X and Y, and search for alternatives instead. Recent research has led to integrating context-adaptable selection criteria with argument-based reasoning mechanisms. In [51], the authors present an approach that expands the capabilities of existing argumentation-based recommender systems.The proposal is based on specifying how to select and use the most appropriate argument comparison criterion based on the selection on the users preferences, giving the possibility of programming by the use of conditional expressions, which argument preference criterion has to be used in each particular situation.

Combining argumentation with trust has received particular attention from the research community in the last years, particularly in connection with reasoning under uncertainty. In [48] the auhors present the design and analysis of a user study which was intended to evaluate the effectiveness of ArgTrust –a decision making tool based in formal argumentation, where the user can assign trust values to evidence– in a collaborative human-agent decision-making task. Empirical results show that users interactions with ArgTrust helped them consider their decisions more carefully than without using the software tool. In [2] an argument-based system is presented that allows an agent to reason about its own beliefs and information received from other sources. An agent's beliefs are of two kinds: beliefs about the environment (e.g. "the window is closed") and beliefs about trusting sources (e,g, agent i trusts agent j). Six basic forms of trust are discussed in the paper, including as well the notion of graded trust (agent i trusts agent j to a certain extent). When attempting to persuade an agent to believe (or disbelieve) an argument, it can be advantageous for the persuader to have a model of the persuadee. In [28] the author introduces a two-dimensional model that accounts for the uncertainty of belief by a persuadee and for the confidence in that uncertainty evaluation. This gives a better modeling for what the user believes/disbelieves as outcomes, and the confidence value is the degree to which the user does indeed hold those outcomes. The framework is also extended with a modelling of the risk of disengagement by the persuadee.

2.3 Computational Persuasion

Persuasion is a human activity that involves a party attempting to influence a person's beliefs, attitudes, intentions, motivations, or behaviors. From its origins on the ancient Greek philosophy persuasion and argumentation theories are intertwined. Aristotle's *Rhetoric* [19] describes the modes of persuasion furnished by the spoken word as of three types: *ethos* –when persuasion is achieved by the speaker's personal authority or credibility; *pathos* –when persuasion comes through the hearers, when the speech appeals to to the audience's emotions; and *logos* –when persuasion comes through the speech itself when the truth (or apparent truth) is proven by means of persuasive *arguments*. Therefore, argumentation can be used as a persuasion device. On the other way round, persuasion is considered as one of the arts included in the interdisciplinary field of argumentation theory, where it has been one of the most studied types of argumentation dialogues [52].

Nowadays, with the advent of artificial intelligence and the exponential growth of the digital data available on the Internet, many theories of human behavior and thinking have been applied to develop advanced information and telecommunication technologies that represent, interact with and advise humans. Among them, persuasion technologies are an interdisciplinary field of research that focuses on the design and development of interactive technologies that can create, maintain or change the thinking and human behavior using persuasion techniques.

Interestingly, although argumentation has a long history of successful applications that provide computational models of persuasion in many domains [41, 42, 49], it does not play a remarkable role in the current persuasion technologies. As pointed out in [26], most current persuasion technologies designed to produce behavioral changes in humans are based on the following approaches:

- Combining questionnaires to find out information about users;
- Providing information to motivate behavior change in users;
- Using computer games that allow users to explore different scenarios with respect to their behavior;
- Providing applications that allow users to record their behavior;
- Providing applications that display messages to encourage users to continue with the improvement of their behavior.

In these systems, the arguments favoring behavior change are just assumed or implicitly provided. The generation and explicit representation of arguments and counter-arguments that can persuade specific people in specific situations is not taken into account. However, both on persuasion in the real world and on applications that promote behavior change, the generation and explicit representation of convincing arguments and arguments against the arguments of other users is crucial. Therefore, persuasion technologies should include computational models of argument that allow:

- Generating and managing arguments and counter-arguments explicitly;
- Creating dialogue protocols controlling the exchange of persuasive arguments between the parties involved; and
- Creating persuasion strategies that make use of a model of the persuadee to select the potentially most effective arguments for each individual at a specific time of the dialogue.

Following with the running example provided in Table 2.1, if the user wants to know the reasons behind the decision of the PA to contact pharmacy *Y*, the system could generate the following textual argument to provide an explanation:

(A) "You need to take pill *B*, so you should call pharmacy *Y*" (from items 3 and 7).

In this way, the PA would be able to appeal the user's logical reasoning and persuade her by proving that its decisions are based on truthful facts (as an instance of the *logos* mode of persuasion). Furthermore, the PA could also include an interface to interact with the user following a persuasion dialogue protocol. Thus, the user could query the PA for taking pill *A*, and the PA could generate a textual argument to explain to the user the reasons why it cannot advise this.

(B) "You cannot call pharmacy *X* to take pill *A*, since today is Saturday and it is closed" (from items 1–2, and 4–6).

In addition, depending on the user's profile and preferences, the PA may follow a particular persuasion strategy. For instance, if the PA's model of the persuadee defines her as a *pragmatic person* who prefers quick and short action recommendations, the PA may only show argument *A* to explain its line of reasoning. Instead, if the PA considers the user to be a *rational person*, it may also show argument *B* to expose all reasons for calling pharmacy *Y* and not *X*.

Currently, the research community has acknowledged the lack of formal argumentation approaches in persuasion technologies and new projects[5] and initiatives have been started to deal with this challenge. On the specific domain of AAL, persuasive PAs can provide understandable justifications for medical diagnosis and health-care recommendations, motivating users' engagement and behavior change. These justifications can come from different sources of knowledge (e.g. scientific texts and clinical practice guidelines [22], randomized controlled trials [30], previous experiences formatted as clinical cases [16], etc.). In the literature, we can find an ambiguous use of the terms 'justification', and 'argument'. The general term 'justification' is used to denote any element that supports the decisions of a system (a PA in our application domain). A justification can be a text, an inspection on the data used by the system to derive a conclusion, a set of rules, etc. Here, we use the term argument to denote

[5] Framework for Computational Persuasion Project: http://www.computationalpersuasion.com/.

the formal representation of the justification for each suggestion or decision (as it is understood by argumentation theory). Thus, arguments are made up of a conclusion and a set of elements that support them, which are typically facts and rules, cases, or schemes, depending on the argumentation formalism used to develop the system.

Several well-known concepts of the argumentation theory have been adopted by the AI community to manage persuasion dialogues in computational systems. Among them, the theory of argumentation schemes has been widely applied. Argumentation schemes represent stereotyped patterns of common reasoning whose instantiation provides an alleged justification for the conclusion drawn from the scheme. The arguments inferred from argumentation schemes adopt the form of a set of general rules by which, given a set of premises, a conclusion can be derived. Different authors have proposed several sets of argumentation schemes, but the work of Walton [53], who presented a set of 25 different argumentation schemes, has been the most used by its simplicity and the fact that his schemes have associated a set of critical questions. If instantiated, these questions can represent potential attacks to the conclusion drawn from the scheme. Therefore, if the opponent asks a critical question, the argument that supports this argumentation scheme remains temporally rebutted until the question is conveniently answered. This characteristic of Walton's argumentation schemes makes them very suitable to reflect reasoning patterns that the system can follow to bring about conclusions and, what is more important, to devise ways of attacking any other alternative conclusions.

In [12], authors presented a persuasive module that has been integrated in a cognitive assistant framework, the iGenda framework [11]. The proposed persuasive module tried to improve user engagement by generating arguments that allow the system to select such health-care activities that best suit to the users' profile. These arguments were based on previous similar cases stored in a case-base, which provided a justification based on the information of the clinical guidelines used to select a specific action. This PA implements a value-based argumentation framework that allows it to automatically schedule the most persuasive activities for each user (those with a higher amount of warranted arguments that support them) by following a case-based reasoning approach. This work covers both recommendation and persuasion areas, but actually persuasion is viewed as an internal process that simulates the interaction between the system and the user.

In a subsequent work [13], authors extended their PA with a new argumentation resource that consists on a set of argumentation schemes that capture the way of reasoning that physicians and caregivers follow to recommend activities to patients. Thus, they provide a way to generate more elaborated arguments (e.g. based on analogy, on the opinion of an expert, or on popular practice) and to determine the relation among arguments (e.g. specifying clearly how an argument can receive attacks). Following a similar approach, although not based on Walton's argumentation schemes, biomedical argumentation schemes are presented as logical programs to be able to automatically mining arguments from clinical guidelines in [22].

The role of argumentation schemes to represent fallacious reasoning in public health was analyzed in [14]. Also, in [5], the author presented ongoing research on testing the effectiveness and usability of argumentation schemes to improve the persuasion power of doctors. The purpose of this system was to enhance elderly diabetes patient's self-management abilities in chronic care. However, those works focus on formal argumentation and linguistics, and do not apply their findings in a computational system. Despite that, they tested argumentation theories with real people in AAL domains, and hence, pave the way for implementing such theories in persuasion technologies developed in PAs.

In addition to argumentation schemes, further work has investigated the general role of argumentation theory to persuade people in the context of medical diagnosis and health care. In [21], authors presented a theory of *informal argumentation* to promote behavior change, focused on the audience's perception of the argument rather than on the soundness of the argument itself. This theory was implemented in a system called *Daphne*, which uses dialectical argumentation for providing healthy nutrition education. In [36], *Portia*, a dialogue system that includes an argumentation module to persuade users to adopt healthy-eating habits, was presented. This system explored different types of argumentation strategies that combine rational arguments and emotional arguments that appeal to the users' expected emotions. Finally, in [24], authors presented an assistive technology that recommends daily activities to its users. The system follows a non-intrusive approach that tracks and monitors individuals' activities and generates recommendations supported by arguments that try to persuade the user towards healthier life styles (e.g. eat healthy, socializing, and maintaining good physical conditions). One of the most interesting contributions of this work is how arguments are translated into pseudo-natural language sentences in order to encourage and give positive feedback to the users.

Although not specifically called PAs after their authors, all these works proposed computational systems that help their users to improve their welfare to some extent by persuading them to change their behavior. Therefore, despite still being reductionist (interfaces to a collection of independent functionalities), they establish a solid background for the application of argumentation-based persuasion technologies in AAL domains.

2.4 Conclusion and Open Issues

Despite not many work in argumentation is specifically devoted to develop PAs in AAL, there is a clear link between these areas. Furthermore, recommendation and persuasion are key functionalities that PAs must have to interact naturally with their users and provide really useful support services to them. In this chapter, we have explored the connection among those technologies and reviewed related work that can play an important role for the development of the next generation of PAs in AAL.

As a summary of our review, Table 2.2 shows how these related work[6] compare in terms of their main research topic, application domain, argumentation formalism,

[6]For the sake of clarity, we have only included in Table 2.2 those reviewed works whose main objective is to provide any type of assistance to their users in an AAL-related domain.

Table 2.2 Summary of the related work

Authors	Research topic	Application domain	Argumentation formalism	Argumentation objective	Architecture	HCI style
Bigi [5]	Persuasion	Diabetes chronic care self-management	Formal Argumentation, Linguistics, Argumentation schemes	Patient education and counseling	n/a	n/a
Costa et al. [12, 13]	Recommendation, Persuasion	AAL, Schedule activities to improve elders' health	Value-based, Case-Based, Argumentation schemes	Recommend most persuasive activity	Modular, Mobile App	Textual alerts
Grasso et al. [21]	Persuasion	Healthy nutrition education	New rhetoric schemes [54]	Engage in persuasive dialogue	Agent-based	Textual utterances
Guerrero et al. [24]	Recommendation, Persuasion	Recommend daily activities to improve health	Rule-based, Possibilistic logic programming	Encourage users, Guide users to healthier behavior	Modular, Mobile App	Pseudo-natural language sentences

(continued)

Table 2.2 (continued)

Authors	Research topic	Application domain	Argumentation formalism	Argumentation objective	Architecture	HCI Style
Hunter and Williams [29]	Recommendation	Aggregation of clinical evidence	Structured arguments based on evidence tables. Argumentation graphs	Decide the best treatment	n/a	n/a
Marcais et al. [34]	Decision making	AAL, Assist elders with Alzheimer disease	Rule-based, Gorgias argumentation framework (preference reasoning and abduction) [31]	Reasoning on pills dosage, Assigning priority to schedule conflicting tasks	Multi-agent system service platform	Textual alerts on Internet-enabled TVs
Mazzotta et al. [36]	Recommendation, Persuasion	Healthy eating habits education	Rule-based, Probabilistic reasoning	Convince users to change eating habits	Agent-based dialogue system	Natural language messages
Muñoz et al. [39]	Decision making	AAL, Monitor elders at home and alert to problems	Rule-based, OWL-DL reasoning	Decide the most reliable sensors information	Multi-agent system	SMS textual alerts, Phone calls

objective of the argumentation process, architecture, and HCI style (i.e. how arguments and information are presented to the user).

As shown in the table, there is a great variability among the argumentation formalism used by the related work, but all have in common their focus on argument generation and evaluation. Hence, argumentation is mainly used to provide a personalized support to users by making decisions or recommendations that suit their profile and objectives, or by generating personalized justifications aimed at convincing them to change their behavior. This establishes a good background to cope with one of the open challenges of PAs; the provision of personalized assistance to users.

Notwithstanding, PAs still have many challenges to overcome in order to declare their success in all of their application domains [10]:

1. To get a real *agency*: they must be able to recognize users' goals, act proactively and collaborate with other people and PAs to accomplish them, and engage in conversations;
2. To provide a *personalized assistance*: they must fit the users' profiles and preferences, interact with their family and social context, solve potential conflicts, take into account their emotional estates; and
3. To deal with *trust, privacy, security and ethical issues*.

Adding up to the provision of personalized assistance, argumentation protocols and argumentation strategies can also elicit actual proactive behavior and natural interaction between PAs and their users. Argumentation research has provided many formal argumentation protocols [37] whose adaptation to AAL domains may provide a formal way to model different types of human-like dialogues between PAs and users. An important open issue here would be to go beyond the usual approach of argumentation systems to assume 'asymmetric dialogues', which simulate users rather than directly interacting with them. This approach is commonly followed due to the difficulties of computational systems to understand the natural language of the user. However, work on computational models of natural argument [43] and contributions on the newly research topic of argument mining [6] provide tools to cope with this challenge. Furthermore, they can also pave the way for improving HCI interfaces, which currently are usually based on the presentation of basic textual pre-compiled messages to the user.

Similarly, recent research on argumentation theory to investigate persuasion strategies in dialogues can advance the current state of the art on PAs and allow them to actually achieve changes in the behavior of their users [27, 44]. Proposals on this area have to pay particular attention on the evaluation of their actual persuasive power and their ability to maintain these behavior changes over time [40].

Finally, interacting with people entails big challenges to build people's trust on the PA and to preserve people's privacy and security. These issues have also been investigated on computational argumentation, and contributions in this domain may inspire new approaches to deal with similar challenges in the PAs for AAL domain. For instance, arguments can be used to justify trust values and support the reputation of a specific PA [50].

Acknowledgements This work was supported by the projects TIN2015-65515-C4-1-R and TIN2014-55206-R of the Spanish government, and by the grant program for the recruitment of doctors for the Spanish system of science and technology (PAID-10–14) of the Universitat Politècnica de València. Authors also acknowledge partial support by PICT-ANPCyT 2014-0624, PIP-CONICET 112-2012010-0487, PGI-UNS 24/N039, and PGI-UNS 24/N040.

References

1. Akbaraly, T., Portet, F., Fustinoni, S., Dartigues, J.-F., Artero, S., Rouaud, O., Touchon, J., Ritchie, K., Berr, C.: Leisure activities and the risk of dementia in the elderly results from the three-city study. Neurology **73**(11), 854–861 (2009)
2. Amgoud, L., Demolombe, R.: An argumentation-based approach for reasoning about trust in information sources. Argum. Comput. **5**(2–3), 191–215 (2014)
3. Atkinson, K., Cerutti, F., McBurney, P., Parsons, S., Rahwan, I.: Special issue on argumentation in multi-agent systems. Argum. Comput. **7**(2–3), 109–112 (2016)
4. Besnard, P., Hunter, A.: Elements of Argumentation. The MIT Press (2008). ISBN 0262026430, 9780262026437
5. Bigi, S.: Healthy reasoning: the role of effective argumentation for enhancing elderly patients' self-management abilities in chronic care. Active Ageing Healthy Living A Hum. Cent. Approach Res. Innov. Sour. Qual. Life **203**, 193 (2014)
6. Cabrio, E., Hirst, G., Villata, S., Wyner, A.: Natural language argumentation: Mining, processing, and reasoning over textual arguments (dagstuhl seminar 16161). In: Dagstuhl Reports, vol. 6. Schloss Dagstuhl-Leibniz-Zentrum fuer Informatik (2016)
7. Chesñevar, C.I., Maguitman, A.G.: Combining argumentation and web search technology: towards a qualitative approach for ranking results. JACIII **9**(1), 53–60 (2005)
8. Chesñevar, C.I., Maguitman, A.G., González, M.P.: Empowering recommendation technologies through argumentation. In Simari and Rahwan [47], p. 403–422. ISBN 978-0-387-98196-3
9. Cocarascu, O., Toni, F.: Argumentation for machine learning: a survey. In: Baroni, P., Gordon, T.F., Scheffler, T., Stede, M. (eds.), Computational Models of Argument—Proceedings of COMMA 2016, Potsdam, Germany, 12-16 September, 2016. volume 287 of Frontiers in Artificial Intelligence and Applications, pp. 219–230. IOS Press (2016). ISBN 978-1-61499-685-9
10. Cohen, P., Cheyer, A., Horvitz, E., El Kaliouby, R., Whittaker, S.: On the future of personal assistants. In: Proceedings of the 2016 CHI Conference Extended Abstracts on Human Factors in Computing Systems, pp. 1032–1037. ACM (2016)
11. Costa, A., Novais, P., Simoes, R.: A caregiver support platform within the scope of an ambient assisted living ecosystem. Sensors **14**(3), 5654–5676 (2014)
12. Costa, A., Heras, S., Palanca, J., Novais, P., Julián, V.: A persuasive cognitive assistant system. In: Ambient Intelligence-Software and Applications–7th International Symposium on Ambient Intelligence (ISAmI 2016), pp. 151–160. Springer International Publishing (2016)
13. Costa, A., Heras, S., Palanca, J., Jordán, J., Novais, P., Julián, V.: Argumentation schemes for events suggestion in an e-health platform. In: XII International Conference on Persuasive Technology XII. To Appear (2017)
14. Cummings, L.: Reasoning and public health: New ways of coping with uncertainty. Springer (2015)
15. Dale, R.: The limits of intelligent personal assistants. Natural Lang. Eng. **21**(02), 325–329 (2015)
16. El-Sappagh, S., Elmogy, M.M.: Medical case based reasoning frameworks: Current developments and future directions. Int. J. Decis. Support Syst. Technol. (IJDSST) **8**(3), 31–62 (2016)
17. EuropeanCommission: The ageing report economic and budgetary projections for the 28 eu member states (2013–2060). Technical report, European Commission, Directorate-General for Economic and Financial Affairs (2015)

18. Ferretti, E., Tamargo, L.H., García, A.J., Errecalde, M.L., Simari, G.R.: An approach to decision making based on dynamic argumentation systems. Artif. Intell. **242**, 107–131 (2017)
19. Freese, J.: Aristotle, The Art of Rhetoric. With Greek text (English translation). Cambridge, Loeb Classical Library/Harvard University Press (1924)
20. Giarratano, J.C., Riley, G.D.: Expert Syst. Princ. Program. Brooks/Cole Publishing Co., Pacific Grove, CA, USA (2005)
21. Grasso, F., Cawsey, A., Jones, R.: Dialectical argumentation to solve conflicts in advice giving: a case study in the promotion of healthy nutrition. Int. J. Hum.-Comput. Stud., 53(6):1077–1115 (2000). ISSN 1071-5819
22. Green, N.: Implementing argumentation schemes as logic programs. In: The 16th Workshop on Computational Models of Natural Argument. CEUR Workshop Proceedings (2016)
23. Grosse, K., González, M.P., Chesñevar, C.I., Maguitman, A.G.: Integrating argumentation and sentiment analysis for mining opinions from twitter. AI Commun. **28**(3), 387–401 (2015)
24. Guerrero, E., Nieves, J.C., Lindgren, H.: An activity-centric argumentation framework for assistive technology aimed at improving health. Argum. Comput., (Preprint), 1–29 (2016)
25. Heras, S., Atkinson, K., Botti, V.J., Grasso, F., Julián, V., McBurney, P.: Research opportunities for argumentation in social networks. Artif. Intell. Rev. **39**(1), 39–62 (2013)
26. Hunter, A.: Opportunities for argument-centric persuasion in behaviour change. In: European Workshop on Logics in Artificial Intelligence, pp. 48–61. Springer (2014)
27. Hunter, A.: Modelling the persuadee in asymmetric argumentation dialogues for persuasion. In: 24th International Joint Conference on Artificial Intelligence, pp. 3055–3061 (2015)
28. Hunter, A.: Two dimensional uncertainty in persuadee modelling in argumentation. In: Kaminka, G.A., Fox, M., Bouquet, P., Hüllermeier, E., Dignum, V., Dignum, F., van Harmelen, F. (eds.), ECAI 2016—22nd European Conference on Artificial Intelligence, 29 August-2 September 2016, The Hague, The Netherlands - Including Prestigious Applications of Artificial Intelligence (PAIS 2016), volume 285 of Frontiers in Artificial Intelligence and Applications, pp. 150–157. IOS Press (2016). ISBN 978-1-61499-671-2
29. Hunter, A., Williams, M.: Aggregation of clinical evidence using argumentation: a tutorial introduction. In: Foundations of Biomedical Knowledge Representation, pp. 317–337. Springer (2015)
30. Jackson, S., Schneider, J.: Argumentation devices in reasoning about health. In: The 16th Workshop on Computational Models of Natural Argument. CEUR Workshop Proceedings (2016)
31. Kakas, A., Moraitis, P.: Argumentation based decision making for autonomous agents. In: Proceedings of the Second International Joint Conference on Autonomous Agents and Multiagent Systems, pp. 883–890. ACM (2003)
32. Konstan, J.A.: Introduction to recommender systems: algorithms and evaluation. ACM Trans. Inf. Syst., 22(1), 1–4 (2004). ISSN 1046-8188
33. Kruse, K.: 15 Secrets Successful People Know About Time Management: The Productivity Habits of 7 Billionaires, 13 Olympic Athletes, 29 Straight-A Students, and 239 Entrepreneurs. The Kruse Group; 1 edition, 1 (2015). ISBN 0985056436
34. Marcais, J., Spanoudakis, N., Moraitis, P.: Using argumentation for ambient assisted living. In: Artificial Intelligence Applications and Innovations, pp. 410–419. Springer (2011)
35. Marsh, S., Dibben, M.R.: Trust, untrust, distrust and mistrust–an exploration of the dark (er) side. In: International Conference on Trust Management, pp. 17–33. Springer (2005)
36. Mazzotta, I., de Rosis, F., Carofiglio, V.: Portia: a user-adapted persuasion system in the healthy-eating domain. IEEE Intell. Syst. **22**(6), 42–51 (2007)
37. McBurney, P., Parsons, S.: Games that agents play: a formal framework for dialogues between autonomous agents. J. Logic Lang. Inf. **11**(3), 315–334 (2002)
38. McCarthy, J.: Some expert systems need common sense. Ann. N.Y. Acad. Sci., 426(1), 129–137 (1984). ISSN 1749-6632
39. Muñoz, A., Augusto, J.C., Villa, A., Botía, J.A.: Design and evaluation of an ambient assisted living system based on an argumentative multi-agent system. Personal Ubiquitous Comput. **15**(4), 377–387 (2011)

40. Nguyen, H., Masthoff, J.: Designing persuasive dialogue systems: using argumentation with care. In: International Conference on Persuasive Technology, pp. 201–212. Springer (2008)
41. Prakken, H.: Formal systems for persuasion dialogue. Knowl. Eng. Rev. **21**(02), 163–188 (2006)
42. Reed, C.: Dialogue frames in agent communication. In: Proceedings of International Conference on Multi Agent Systems, pp. 246–253. IEEE (1998)
43. Reed, C., Grasso, F.: Recent advances in computational models of natural argument. Int. J. Intell. Syst. **22**(1), 1–15 (2007)
44. Rosenfeld, A., Kraus, S.: Strategical argumentative agent for human persuasion. In: ECAI 2016: 22nd European Conference on Artificial Intelligence, 29 August-2 September 2016, The Hague, The Netherlands-Including Prestigious Applications of Artificial Intelligence (PAIS 2016), vol. 285, p. 320. IOS Press (2016)
45. Sandvig, J.J., Mobasher, B., Burke, R.D.: A survey of collaborative recommendation and the robustness of model-based algorithms. IEEE Data Eng. Bull. **31**(2), 3–13 (2008)
46. Schlesinger, F., Ferretti, E., Errecalde, M., Aguirre, G.: An argumentation-based BDI personal assistant. In: García-Pedrajas, N., Herrera, F., Fyfe, C., Benítez, J.M., Ali, M. (eds.), Trends in Applied Intelligent Systems - 23rd International Conference on Industrial Engineering and Other Applications of Applied Intelligent Systems, IEA/AIE 2010, Cordoba, Spain, June 1-4, 2010, Proceedings, Part I, vol. 6096 of Lecture Notes in Computer Science, pp. 701–710. Springer (2010). ISBN 978-3-642-13021-2
47. Simari, G.R., Rahwan, I. (eds.): Argumentation in Artificial Intelligence. Springer (2009). ISBN:978-0-387-98196-3
48. Sklar, E.I., Parsons, S., Li, Z., Salvit, J., Perumal, S., Wall, H., Mangels, J.A.: Evaluation of a trust-modulated argumentation-based interactive decision-making tool. Auton. Agents Multi-Agent Syst. **30**(1), 136–173 (2016)
49. Sycara, K.P.: Persuasive argumentation in negotiation. Theo. Decis. **28**(3), 203–242 (1990)
50. Tang, Y., Cai, K., McBurney, P., Sklar, E., Parsons, S.: Using argumentation to reason about trust and belief. J. Logic Comput., exr038 (2011)
51. Teze, J.C., Gottifredi, S., García, A.J., Simari, G.R.: Improving argumentation-based recommender systems through context-adaptable selection criteria. Expert Syst. Appl. **42**(21), 8243–8258 (2015)
52. Walton, D., Krabbe, E.C.: Commitment in dialogue: Basic concepts of interpersonal reasoning. SUNY Press (1995)
53. Walton, D., Reed, C., Macagno, F.: Argumentation Schemes. Cambridge University Press (2008)
54. Warnick, B., Kline, S.L.: The new rhetorics argument schemes, a rhetorical view of practical reasoning. Argum. Advocacy, 29(1), 1–15 (1992)

Chapter 3
Kidney Care—A Personal Assistant Assessment

Bia Martins, Joao Rei, Miguel Braga, Antonio Abelha, Henrique Vicente, Joao Neves and Jose Neves

Abstract A cognitive disability is a medical condition that, despite all technological progress, still does not have a cure, i.e., there are cases where the physician may use medication, but the only purpose is to decrease the progression of the disease, not its cure. This is the case in many situations, and in particular in kidney illnesses, which have a dominant impact on a person well being, i.e., the assistance to an individual to whom was diagnosed cognitive disabilities is essential, where the location of the individual is not decisive or important. Hence, the presence of a Personal Assistance Service can become a cornerstone in achieving independence and quality of life. Therefore, the objective of this work is to present an intelligent system aimed at an endless individuals monitoring and alerting system, based on a Logical Programming approach to Knowledge Representation and Reasoning, and centre on RapidMiner, a software platform that provides an integrated environment

B. Martins · J. Rei · M. Braga · A. Abelha · H. Vicente · J. Neves (✉)
Centro Algoritmi, Universidade do Minho, Braga, Portugal
e-mail: jneves@di.uminho.pt

B. Martins
e-mail: a69990@alunos.uminho.pt

J. Rei
e-mail: a68381@alunos.uminho.pt

M. Braga
e-mail: a68376@alunos.uminho.pt

A. Abelha
e-mail: abelha@di.uminho.pt

H. Vicente
e-mail: hvicente@uevora.pt

H. Vicente
Departamento de Química, Escola de Ciências e Tecnologia, Universidade de Évora, Évora, Portugal

J. Neves
Mediclinic Arabian Ranches, 282602, Dubai, United Arab Emirates
e-mail: joaocpneves@gmail.com

A. Costa et al. (eds.), *Personal Assistants: Emerging Computational Technologies*, Intelligent Systems Reference Library 132, DOI 10.1007/978-3-319-62530-0_3

for machine learning, predictive analysis or application development and deployment. It undergoes a Case Based approach to computing that tracks patient's performance, learn and deliver content when it is needed, and assures that patient's key information is changed into the indispensable ongoing knowledge.

Keywords Personal Assistants Artificial Intelligence Intelligent Systems · Logic Programming · Knowledge Representation and Reasoning · Case Based Reasoning

3.1 Introduction

Kidneys play an important role in the human organism, once are involved in a wide variety of processes in the human body. Clinical assessment of this organ's function is common in routine medical practice and essential for assessing overall health and, more specifically, for detecting, evaluating and monitoring acute and chronic diseases. Indeed, in healthcare there has been a growing interest and investment in various forms of constant individuals monitoring. The increasing of average life expectation and, consequently, the costs in healthcare due to the population in general, and the elderly in particular, are the motivation for this investment. There are several ways to evaluate kidney function, but the Glomerular Filtration Rate (GFR) is considered the best overall index of kidney function in health and disease. GFR measures the clearance of a substance that is freely filtered through the kidney's glomerular capillary wall in a certain period of time. Although this is considered the greatest method to assess kidney function, it is not easy to measure GFR in clinical practice, once its rate is estimated from equations using serum creatinine, age, race, gender or body size [1, 2]. On the other hand, creatinine fulfils almost all the requirements for a perfect filtration marker, once it is not a protein bound, it is freely filtered and not metabolized by the kidney, and it is physiologically inert. If one's kidney is not functioning properly, an increased level of creatinine may accumulate in one's blood. A serum creatinine test measures the level of creatinine in the blood and provides an estimate of how well the kidney filter (glomerular filtration rate). These reasons explain why is creatinine often considered the main substance to calculate GFR. Knowing both its values in the blood and urine, it is possible to calculate the creatinine clearance and, as a consequence, to evaluate a patient's kidney function. However, creatinine concentration in the blood is not, by itself, enough to predict kidney's state, once it varies according characteristics of the patient, namely age, body mass index and gender. For example, for both men and women, normal values of creatinine clearance tend to be lower for older people [3].

Therefore, all the information available is relevant in order to monitor a patient's kidney function. This is the main reason for one's approach to handle the problem, i.e., through *Case Based Reasoning* (*CBR*) it is possible to classify new unlabelled cases using past cases' knowledge. Indeed, one of the major catalysts for change is going to be *Artificial Intelligence* (*AI*), where *CBR* is assumed to be one of the most promising *AI* based techniques to mature intelligent grounded decision support systems [4, 5]. *AI* in healthcare and medicine may organize the patient routes, optimize treatment plans and also provide physicians with literally all the information they need to make a good decision.

3.2 Knowledge Representation and Reasoning

In specific judgments the available data, information or knowledge is not always exact in the sense that it can be estimated values, probabilistic measures, or degrees of uncertainty. Furthermore, knowledge and belief are generally incomplete, self-contradictory, or even error sensitive, being desirable to use formal tools to deal with the problems that arise from the use of these types of information [6, 7]. Indeed, many approaches to Knowledge Representation and Reasoning have been proposed using the *Logic Programming* (*LP*) epitome, namely in the area of *Model Theory* [8, 9] and *Proof Theory* [6, 7]. In the present work the proof theoretical approach is followed in terms of an extension to *LP*. An *Extended Logic Program* is a finite set of clauses as shown in Program 3.1:

The first clause of Program 3.1 stand for predicate's closure, "," denotes "*logical and*", while "*?*" is a domain atom denoting "*falsity*". The "p_i, q_j, and p" are classical ground literals, i.e., either positive atoms or atoms preceded by the classical negation sign "¬" [6]. Indeed, "¬" stands for a "*strong declaration*" that speaks for itself, and "*not*" denotes "*negation-by-failure*", or in other words, a flop in proving a given statement, once it was not declared explicitly. Every program is also associated with a set of "*abducibles*" [8, 9], given here in the form of "*exceptions*"

Program 3.1 An example of a generic *Extended Logic Program*

$$
\begin{array}{l}
\{ \\
\quad \neg\, p \leftarrow not\ p, not\ exception_p \\
\quad p \leftarrow p_1, \cdots, p_n, not\ q_1, \cdots, not\ q_m \\
\quad ?\,(p_1, \cdots, p_n, not\ q_1, \cdots, not\ q_m)(n, m \geq 0) \\
\quad exception_{p_1} \\
\quad \ldots \\
\quad exception_{p_j}\ (0 \leq j \leq k),\ being\ k\ an\ integer \\
\} :: scoring_{value}
\end{array}
$$

to the extensions of the predicates that make the overall program, i.e., clauses of the form:

$$exception_{p_1}, \ldots, exception_{p_j} (0 \leq j \leq k), being\ k\ an\ integer \quad (1)$$

that stands for data, information or knowledge that cannot be ruled out. On the other hand, clauses of the type:

$$?\,(p_1, \ldots, p_n, not\ q_1, \ldots, not\ q_m)(n, m \geq 0) \quad (2)$$

also named invariants or restrictions, allow one to set the context under which the universe of discourse has to be understood. The term "$scoring_{value}$" stands for the relative weight of the extension of a specific "*predicate*" with respect to the extensions of the peers ones that make the inclusive or global program.

3.2.1 Quantitative Knowledge

In order to set the proposed approach to knowledge representation, two metrics will be set, namely the *Quality-of-Information* (*QoI*) of a logic program that will be understood as a mathematical function that will return a truth-value ranging between 0 and 1 [10, 11], once it is fed with the extension of a given predicate, i.e., $QoI_i = 1$ when the information is known (positive) or false (negative) and $QoI_i = 0$ if the information is unknown. For situations where the extensions of the predicates that make the program also include abducible sets, its terms (or clauses) present a $QoI_i\ \epsilon]0, 1[$, in the form:

$$QoI_i = 1/Card \quad (3)$$

if the abducible set for predicates i and j satisfy the invariant:

$$?\left(\left(exception_{p_i}; exception_{p_j}\right), \neg\left(exception_{p_i}; exception_{p_j}\right)\right) \quad (4)$$

where ";" denotes the *logical or* and *Card* stands for set cardinality, being $i \neq j$ and $i, j \geq 1$. Conversely, if there is no constraint on the possible combinations among the abducible clauses, the clauses cardinality (K) will be given by $C_1^{Card} + \cdots + C_{Card}^{Card}$, being the *QoI* acknowledged as:

$$QoI_{i_{1 \leq i \leq Card}} = 1/C_1^{Card}, \cdots, 1/C_{Card}^{Card} \quad (5)$$

where C_{Card}^{Card} is a card-combination subset, with *Card* elements. For example, consider the logic program depicted in Program 3.2:

In Program 3.2 $\perp$ denotes a null value of the type *unknown*. It is now possible to split the *abducible* or *exception* set into the admissible clauses or terms and evaluate

$\{$

$\neg f_1(x,y,z) \leftarrow not\ f_1(x,y,z), not\ exception_{f_1}(x,y,z)$

$f_1\underbrace{(2.8,\ [3,5],\ \perp)}_{attribute`s\ values}$

$\underbrace{[0,5]\ [10,20][5,10]}_{attribute`s\ domains}$

$exception_{f_1}(\perp,\ 2,\ [1,5])$

$\ldots$

$exception_{f_k}([2,4],\ \perp,\ 8)$

$\}$:: 1 *(once the universe of discourse is set in terms of the extension of one predicate*

Program 3.2 An *Extended Logic Program* with three attributes *x*, *y* and *z*

their *QoIs*. A pictorial view of this process, in general terms, is given below as a pie chart (Fig. 3.1).

However, a term's *QoI* also depends on their attribute's *QoI*. In order to evaluate this metric, look to Fig. 3.2, where the segment with bounds 0 and 1 stands for every attribute domain, i.e., all the attributes range in the interval [0, 1]. [A, B] denotes the range where the unknown attributes values for a given predicate may occur. Therefore, the *QoI* of each attribute's clause is calculated using:

$$QoI_{attribute_i} = 1 - \|A - B\| \tag{6}$$

where ||A–B|| stands for the modulus of the arithmetic difference between *A* and *B*. Thus, in Fig. 3.3 is showed the *QoI*'s values for the abducible set for $predicate_i$.

Under this setting, a new evaluation factor has to be considered, which will be denoted as *DoC*, that stands for one's confidence that the argument values or attributes of the terms that make the extension of a given predicate, having into consideration their domains, fit into a given interval [12]. The *DoC* is evaluated as shown in Fig. 3.4 and computed using $DoC = \sqrt{1 - \Delta l^2}$, where Δl stands for the argument interval length, which was set in the interval [0, 1]. Thus, the universe of discourse is engendered according to the information presented in the extensions of such predicates, according to productions of the type:

$$\begin{aligned} predicate_i - \bigcup_{1 \le j \le m} clause_j(((A_{x_1}, B_{x_1})(QoI_{x_1}, DoC_{x_1})), \cdots \\ \cdots, ((A_{x_n}, B_{x_n})(QoI_{x_m}, DoC_{x_m}))) :: QoI_j :: DoC_j \end{aligned} \tag{7}$$

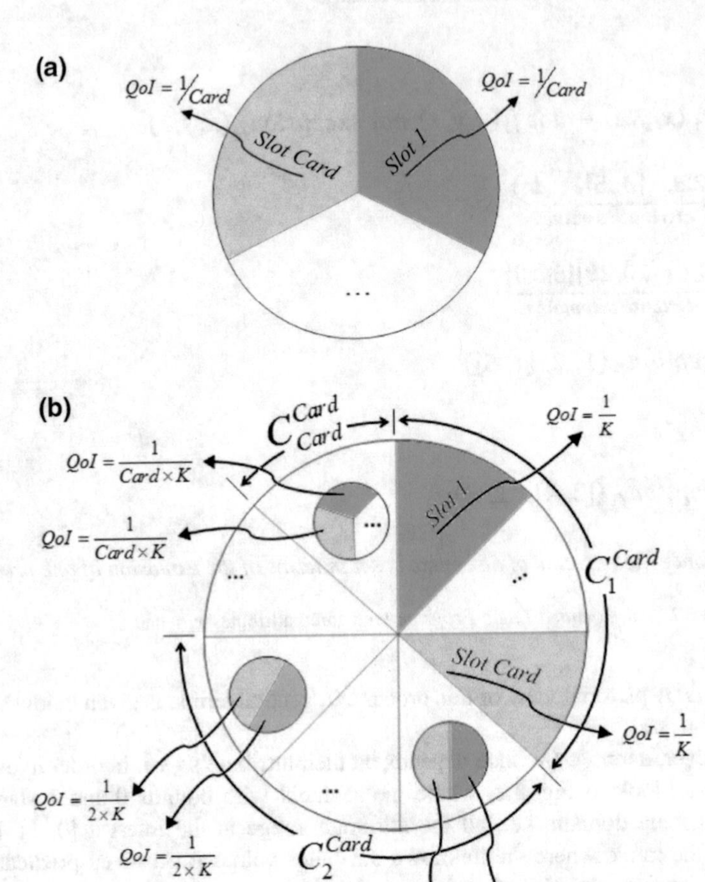

Fig. 3.1 *QoI*'s values for the abducible set for $predicate_i$ with (a) and without (b) constraints on the possible combinations among the abducible clauses

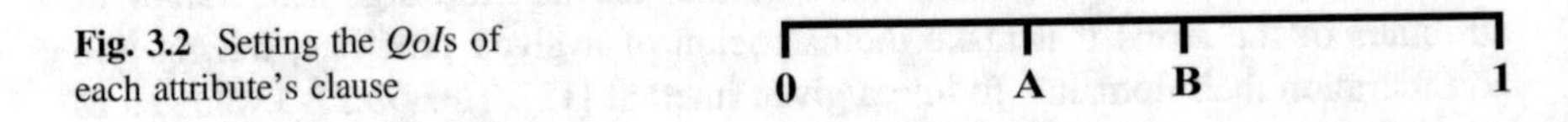

Fig. 3.2 Setting the *QoI*s of each attribute's clause

where ∪, m and (A_{xj}, B_{xj}) stand for, respectively, for set union, the cardinality of $predicate_i$ extension and the extremes of the interval where attribute $attribute_i$ may, in principle, be situated.

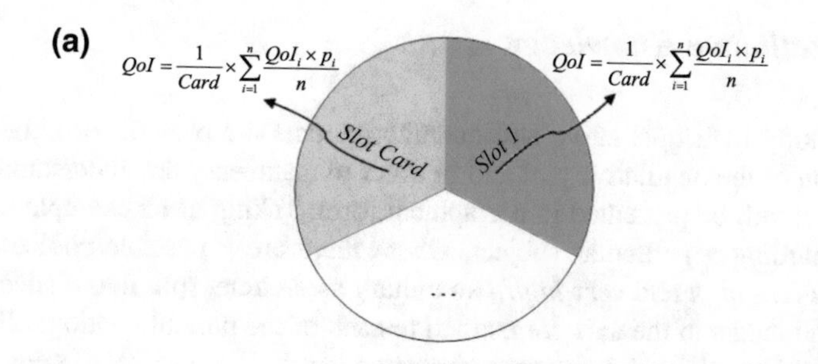

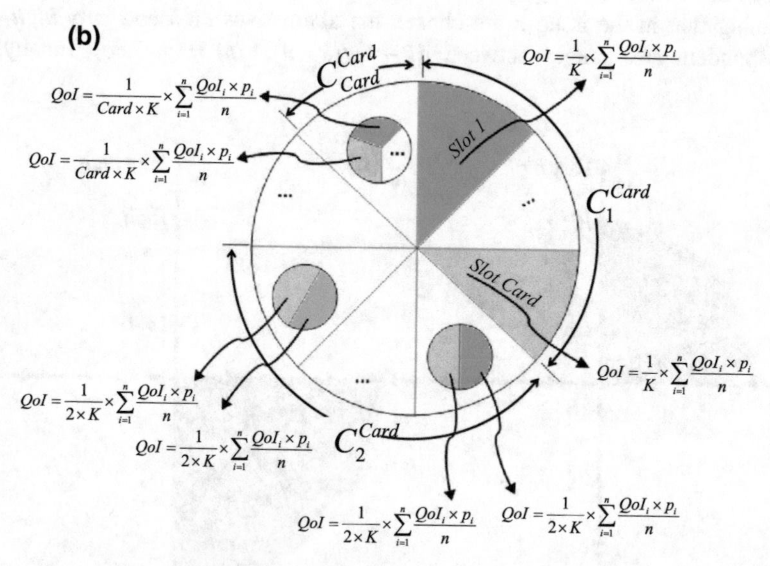

Fig. 3.3 *QoI*'s values for the abducible set for *predicate*$_i$ with (a) and without (b) constraints on the possible combinations among the abducible clauses. $\sum_{i=1}^{n} (QoI_i \times p_i)/n$ denotes the *QoI*'s average of the attributes of each clause (or term) that sets the extension of the *predicate* under analysis. n and p_i stand for, respectively, for the attribute's cardinality and the relative weight of attribute p_i with respect to its peers, being $\sum_{i=1}^{n} p_i = 1$

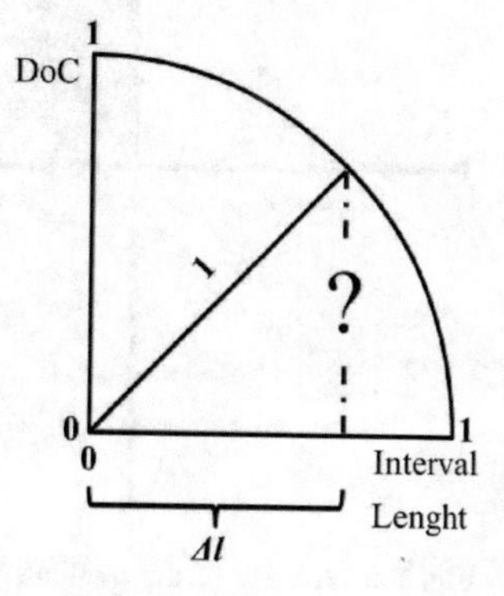

Fig. 3.4 Evaluation of the attributes' *Degree of Confidence*

3.2.2 Qualitative Knowledge

In present study both qualitative and quantitative data are present. Aiming at the quantification of the qualitative part and in order to make easy the understanding of the process, it will be presented in a graphical form. Taking as an example a set of n issues regarding a particular subject, where there are k possible choices (e.g., *absence*, *low*, …, *high* and *very high*), an unitary area circle, split into n slices, was itemized. The marks in the axis correspond to each of the possible options. Thus, if the answer to issue 1 is *high* the area correspondent is $(k-1)/(k \times n)$ (Fig. 3.5a). Assuming that in the issue 2 are chosen the alternatives *high* and *very high*, the correspondent area ranges between $[(k-1)/(k \times n), 1/n]$ (Fig. 3.5b). Finally, in

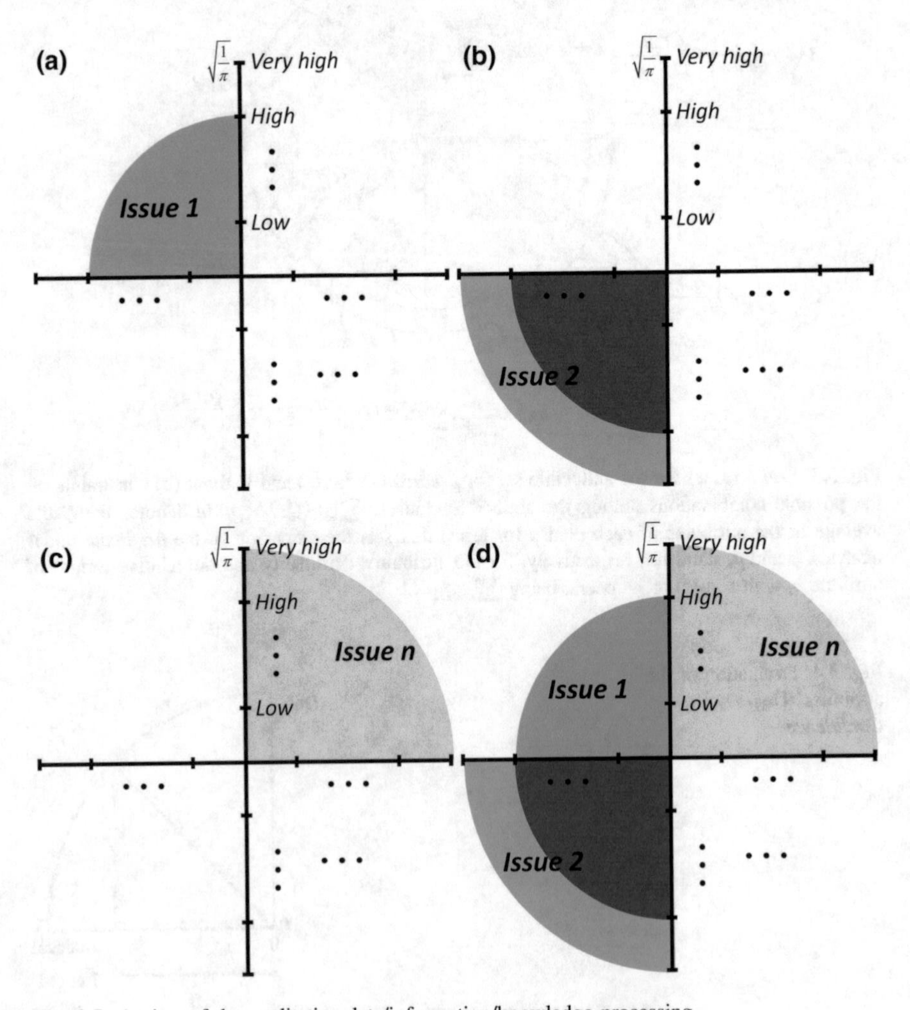

Fig. 3.5 A view of the qualitative data/information/knowledge processing

issue n if no alternative is ticked, all the hypotheses should be considered and the area varies in the interval $[0, 1/n]$ (Fig. 3.5c). The total area is the sum of the partial ones (Fig. 3.5d).

In some cases, similar responses to different issues have opposing impact in the subject in consideration. For example, the assessment of healthy lifestyle includes issues like physical exercise practices and smoking status. The response *high* to the former issue has a positive contribution for healthy lifestyle, while the same response to smoking status has a negative one. Thus, the contribution of the items with negative impact on the subject in analysis is set as $1/n$ minus the correspondent area, i.e., $(1/n - (k-1)/(k \times n)) = 1/(k \times n)$ for issue 1, $[0, 1/(k \times n)]$ for issue 2 and $[0, 1/n]$ for issue 3.

3.3 A Case Based Approach to Problem Solving

The *Case Based* (*CB*) methodology for problem solving presents a whole different approach when compared to others. Indeed, it stands for an act of finding and justifying a solution to a given problem based on the consideration of similar past ones, by reprocessing and/or adapting its information or knowledge [4, 5]. In *CB* the *cases* are stored in a *Case Base*, and those cases that are similar (or close) to a new one are used in the problem solving process. The typical *CB* cycle presents the mechanism that should be followed, where the former stage entails an initial description of the problem. The new case is defined and it is used to retrieve one or more cases from the *CB*. On the other hand, despite promising results, current *CB* systems are neither complete nor adaptable enough for all domains. In fact, there are circumstances where the user can not choose the method (s) of similarity (s) and is required to follow a predefined (s), even if they do not meet their needs. Moreover, in real problems, access to all necessary information is not always possible, since existing *CB* systems have limitations related to the ability to deal explicitly with unknown, incomplete or even contradictory information. Neves et al. [13, 14] induced a different *CB* cycle (Fig. 3.6), which takes into account the case's *QoI* and *DoC* metrics. It also contemplates a case optimization process, when the retrieved cases do not meet the terms under which a given problem has to be handled (e.g., the expected *DoC* on a forecast has not been reached) [13, 14]. Here, the optimization process may use *Artificial Neural Networks* [15], *Particle Swarm Optimization* [16] or *Genetic Algorithms* [7], generating a set of new cases that will be used in the problem solving process and are conform to the invariant:

$$\bigcap_{i=1}^{n} (B_i, E_i) \neq \varnothing \tag{8}$$

It states that the intersection of the attribute's values ranges for cases' set that make the *Case Base* or their optimized counterparts ($\boldsymbol{B_i}$) (being n its cardinality), and the ones that were object of a process of optimization ($\boldsymbol{E_i}$), cannot be empty, i.e., then need to overlap in a certain degree.

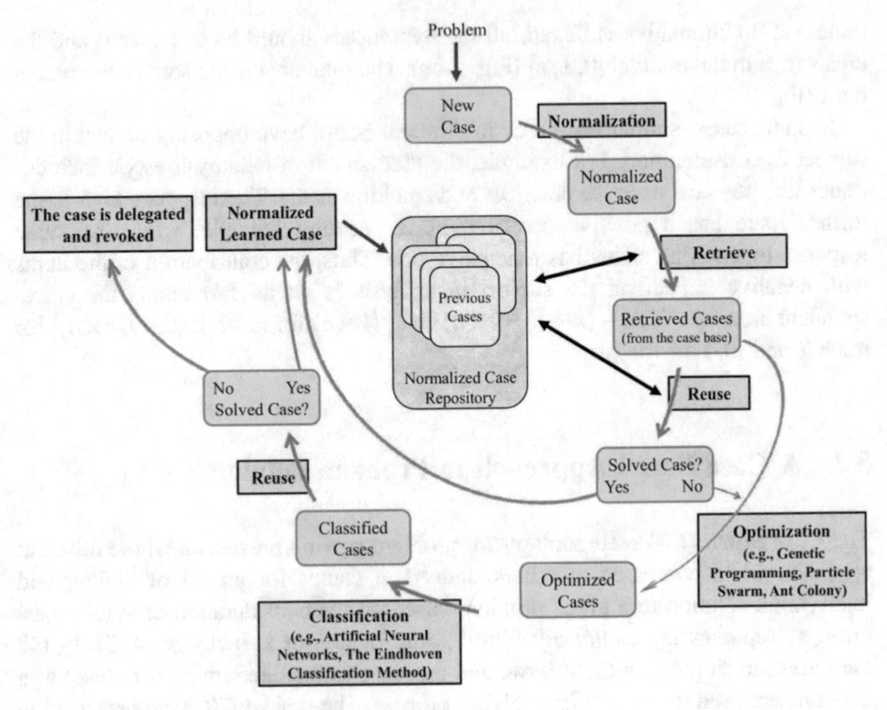

Fig. 3.6 The updated view of the *CB* cycle proposed by Neves et al. [13, 14]

There are examples in the literature on the use of *CBR* as a methodology for problem solving in medicine. Diverse connoisseurs have examined more than thirty *CBR* systems/projects [17, 18], revealing that *CBR* has been widely employed in the medical field, including diagnosis, therapeutics or management. Therefore, and in order to apply this technique to *Personal Assistants* it was mandatory to reduce processing time to find similar cases and to bypass adaptation issues. The former question was accomplished as a consequence of the way information is represented (see Sect. 3.2, above). The latter was obtained through the use of soft computing approaches in the manner just mentioned above. It should also be stated that when listing a case, quantitative information may be intermingled with qualitative one, which may be of type unknown, incomplete or even self-contradictory, a fundamental limitation in classical *CBR*.

3.4 System's Architecture

An intelligent *Personal Assistant* stands for a software agent that may act to the service of people or other agents, working on a base of inputs either from users, awareness of a given location, or from online sources, i.e., a technology that is

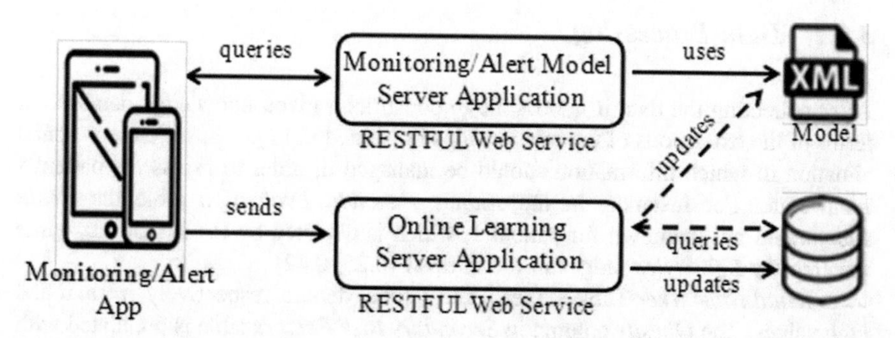

Fig. 3.7 The System's Architecture [19, 21]

enabled and may be understood as a blend of mobile devices, *Application Programming Interfaces* (*APIs*), or mobile apps. Examples of tasks in the health sector may include schedule and personal management or kidney or heart rate assessment, just to name a few [19]. On the other hand, existing tools for monitoring are mostly available as web applications, a reason why the service architecture being used is the one shown below, which was preferred for being light-weight, easy to access and scalable (Fig. 3.7).

The web services were developed in *Java* with the *Java API* for *RESTful Web Services* (*JAX-RS*) (*Oracle 2014*). The data is sent over the *HTML POST* method when the health care professional submit the patient features on the monitoring App. The response, from *RapidMiner* [20], is returned on a *JSON* format. A custom implementation in the logic programming language *PROLOG*, with the same purpose, was also developed and tested. It makes up a mobile ad hoc network, consisting of mobile nodes that use wireless interfaces to send packet data. The nodes act as both routers and hosts, forwarding packets to other nodes as well as run user applications. Indeed, network functions such as routing, address allocation, authentication, and authorization where designed to cope with a dynamic and volatile network topology.

3.5 Case Study

With the objective of developing a predictive model to evaluate the patient's renal function, a database was established based on the medical records of patients from a major health institution in the north of Portugal. In this study 273 cases were considered with an age average of 57.8 years, ranging from 34 to 95 years old. The gender distribution was 41.9 and 58.1% for male and female, respectively.

3.5.1 Data Processing

After collecting the data it is possible to construct a given knowledge database in terms of the extensions of the relationships illustrated in Fig. 3.8, which represent a situation in which information should be managed in order to assess the patient`s renal state. For instance, in the *Kidney Function Evaluation* table the *Renal Biomarkers* are unknown for patient 1, which is depicted by the symbol ⊥, while the *Healthy Life Styles* ranges in the interval [0.25, 0.42].

In *Renal Biomarker* Table 0 (zero) and 1 (one) denote, respectively, *normal* and *high* values. The *Obesity* column in *Secondary Risk Factors* table is populated with 0 (zero), 1 (one) or 2 (two) according to patient' Body Mass Index (BMI). Thus, 0 (zero) denotes $BMI < 25$; 1 (one) stands for a *BMI* ranging in interval [25, 30[; and 2 (two) denotes a $BMI \geq 30$. The remaining columns of *Secondary Risk Factors* and the *Primary Risk Factors* tables are populated with 0 (zero), 1 (one) denoting, respectively, *absence* and *presence*. To set the information present in the *Healthy Lifestyle* table, the procedures described above were followed.

In *Gender* column of the *Kidney Function Evaluation* Table 0 (zero) and 1 (one) stand, respectively, for *female* and *male*, while in *Family Story* column denote, respectively, *no* and *yes*. The values presented in the *Renal Biomarkers* (*RB*), *Primary Risk Factors* (*PRF*) and *Secondary Risk Factors* (*SRF*) columns are the sum of the correspondent tables, ranging between [0, 4], [0, 2] and [0, 7], respectively.

Applying the algorithm presented in [12] to the table or relation's fields that make the knowledge base for kidney function evaluation (Fig. 3.8), and looking to

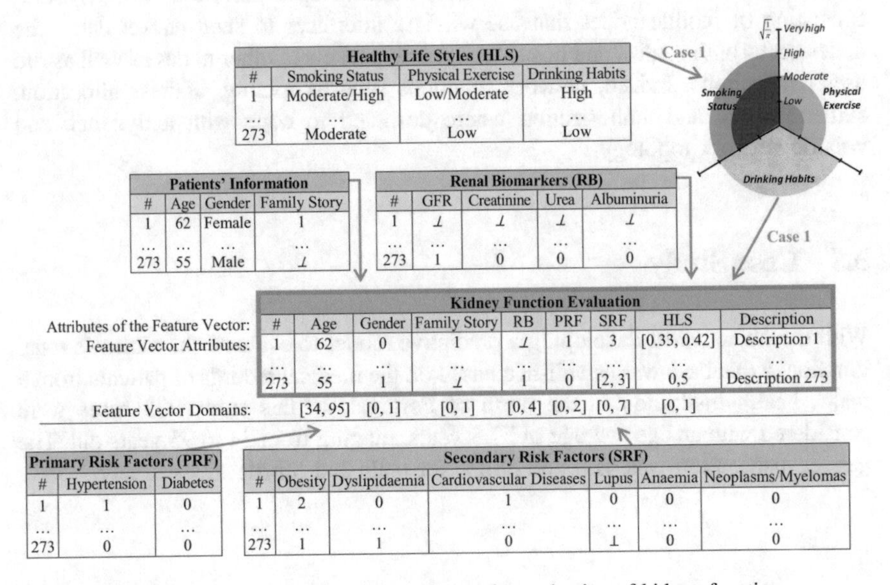
Healthy Life Styles (HLS)

#	Smoking Status	Physical Exercise	Drinking Habits
1	Moderate/High	Low/Moderate	High
…	…	…	…
273	Moderate	Low	Low

Patients' Information

#	Age	Gender	Family Story
1	62	Female	1
…	…	…	…
273	55	Male	⊥

Renal Biomarkers (RB)

#	GFR	Creatinine	Urea	Albuminuria
1	⊥	⊥	⊥	⊥
…	…	…	…	…
273	1	0	0	0

Kidney Function Evaluation

	#	Age	Gender	Family Story	RB	PRF	SRF	HLS	Description
Feature Vector Attributes:	1	62	0	1	⊥	1	3	[0.33, 0.42]	Description 1
	…	…	…	…	…	…	…	…	…
	273	55	1	⊥	1	0	[2, 3]	0,5	Description 273
Feature Vector Domains:		[34, 95]	[0, 1]	[0, 1]	[0, 4]	[0, 2]	[0, 7]	[0, 1]	

Primary Risk Factors (PRF)

#	Hypertension	Diabetes
1	1	0
…	…	…
273	0	0

Secondary Risk Factors (SRF)

#	Obesity	Dyslipidaemia	Cardiovascular Diseases	Lupus	Anaemia	Neoplasms/Myelomas
1	2	0	1	0	0	0
…	…	…	…	…	…	…
273	1	1	0	⊥	0	0

Fig. 3.8 A fragment of the knowledge base aiming the evaluation of kidney function

the *DoCs* values obtained as described before, it is possible to set the arguments of the predicate ***evaluation_of_kidney_function*** ($eval_{kf}$) referred to below, whose extensions denote the objective function regarding the problem under analyse:

$$eval_{kf}\!: Age, G_{ender}, F_{amily}S_{tory}, R_{enal}B_{iomarkers}, P_{rimary}R_{isk} \\ F_{actors}, S_{econdary}R_{isk}F_{actors}, H_{ealthy}L_{ife}S_{tyle} \rightarrow \{0, 1\} \tag{9}$$

where 0 (zero) and 1 (one) denote, respectively, the truth values *false* and *true*.

The algorithm presented in [12] encompasses different phases. In the former one the clauses or terms that make extension of the predicate under study are established. In the subsequent stage the arguments of each clause are set as continuous intervals. In a third step the boundaries of the attributes intervals are set in the interval [0, 1], according to the expression $(Y - Y_{min})/(Y_{max} - Y_{min})$, where the Y_s stand for themselves. Finally, the *DoC* is evaluated.

As an example considers the Program 3.3, regarding a term (patient) that presents the feature vector $Age = 57$, $G_{ender} = 0$, $F_{amily}S_{tory} = \bot$, $R_{enal}B_{iomarkers} = 1$, $P_{rimary}R_{isk}\ F_{actors} = 1$, $S_{econdary}\ R_{isk}F_{actors} = [0, 2]$, $H_{ealthy}L_{ife}S_{tyle} = [0, 0.33]$, one may obtain:

3.5.2 The CBR Approach to Computing

The framework presented previously shows how the information comes together. In this section, a soft computing approach was set to model the universe of discourse, where the computational part is based on a *CB* approach to processing. Contrasting with other problem solving methodologies (e.g., *Decision Trees* or *Artificial Neural Networks*), in *CBR* relatively little work is done offline [22]. Undeniably, in almost all the situations the work is performed at query time. The main difference between the new method [13, 14] and the typical *CB* one [4, 5] relies on the fact that not only all the cases have their arguments set in the interval [0, 1], but it also caters for the handling of incomplete, unknown, or even self-contradictory data or knowledge. Thus, the *Case Base* is given in terms of the following pattern:

$$Case = \{ < Raw_{data}, Normalized_{data}, Description_{data} > \} \tag{10}$$

When confronted with a new case, the system is able to retrieve all cases that meet such a structure and optimize such a population, having in consideration that the cases retrieved from the *Case-base* must satisfy the invariant present in Eq. (8), in order to ensure that the intersection of the attributes range in the cases that make the *Case Base* repository or their optimized counterparts, and the equals in the new case cannot be empty. Having this in mind, the algorithm presented above is applied to a new case, that presents feature vector $Age = 75$, $G_{ender} = 1$, $F_{amily}S_{tory} = 0$, $R_{enal}B_{iomarkers} = \bot$, $P_{rimary}R_{isk}F_{actors} = [0,\ 1]$, $S_{econdary}R_{isk}F_{actors} = 2$,

***Begin** (DoCs evaluation)*
The predicate's extension that sets the Universe-of-Discourse for the term under observation is fixed
{

$$\neg\, eval_{kf}\left(\left((A_{Age}, B_{Age})(QoI_{Age}, DoC_{Age})\right), \cdots \right.$$
$$\left. \cdots, \left((A_{HLS}, B_{HLS})(QoI_{HLS}, DoC_{HLS})\right)\right)$$
$$\leftarrow not\; eval_{kf}\left(\left((A_{Age}, B_{Age})(QoI_{Age}, DoC_{Age})\right), \cdots \right.$$
$$\left. \cdots, \left((A_{HLS}, B_{HLS})(QoI_{HLS}, DoC_{HLS})\right)\right)$$
$$eval_{kf}\left(\left((57, 57)(1_{[57,57]}, DoC_{[57,57]})\right), \cdots \right.$$
$$\left. \cdots, \left((0, 0.33)(1_{[0,0.33]}, DoC_{[0,0.33]})\right)\right) :: 1 :: DoC$$
$$\underbrace{[34, 95] \quad \cdots \quad [0, 1]}_{attribute`s\ domains}$$

}:: 1

The attribute's boundaries are set to the interval [0, 1], according to a normalization process that uses the expression $(Y - Y_{min})/(Y_{max} - Y_{min})$
{

$$\neg\, eval_{kf}\left(\left((A_{Age}, B_{Age})(QoI_{Age}, DoC_{Age})\right), \cdots \right.$$
$$\left. \cdots, \left((A_{HLS}, B_{HLS})(QoI_{HLS}, DoC_{HLS})\right)\right)$$
$$\leftarrow not\; eval_{kf}\left(\left((A_{Age}, B_{Age})(QoI_{Age}, DoC_{Age})\right), \cdots \right.$$
$$\left. \cdots, \left((A_{HLS}, B_{HLS})(QoI_{HLS}, DoC_{HLS})\right)\right)$$
$$eval_{kf}\left(\left((0.38, 0.38)(1_{[0.38,0.38]}, DoC_{[0.38,0.38]})\right), \cdots \right.$$
$$\left. \cdots, \left((0, 0.33)(1_{[0,0.33]}, DoC_{[0,0.33]})\right)\right) :: 1 :: DoC$$
$$\underbrace{[0, 1] \quad \cdots \quad [0, 1]}_{attribute`s\ domains}$$

}:: 1

Program 3.3 An *Extended Logic Program* in order to evaluate the *DoC* of a term that presents the feature vector $Age = 57$, $G_{ender} = 0$, $F_{amily}S_{tory} =$, $R_{enal}B_{iomarkers} = 1$, $P_{rimary}R_{isk}F_{actors} = 1$, $S_{econdary}\ R_{isk}F_{actors} = [0, 2]$, $H_{ealthy}L_{ife}S_{tyle} = [0, 0.33]$

The DoC's values are evaluated

{

$$\neg\, eval_{kf}\left(\left((A_{Age}, B_{Age})(QoI_{Age}, DoC_{Age})\right), \cdots \right.$$
$$\left. \cdots, \left((A_{HLS}, B_{HLS})(QoI_{HLS}, DoC_{HLS})\right)\right)$$
$$\leftarrow not\ eval_{kf}\left(\left((A_{Age}, B_{Age})(QoI_{Age}, DoC_{Age})\right), \cdots \right.$$
$$\left. \cdots, \left((A_{HLS}, B_{HLS})(QoI_{HLS}, DoC_{HLS})\right)\right)$$

$$eval_{kf}\underbrace{\left(((0.38, 0.38)(1, 1)), \cdots, ((0, 0.33)(1, 0.94))\right)}_{\substack{attribute`s\ values\ ranges\ once\ normalized\ and \\ respective\ QoI\ and\ DoC\ values}} :: 1 :: 0.84$$

$$\underbrace{[0, 1] \qquad \cdots \qquad [0, 1]}_{attribute`s\ domains\ once\ normalized}$$

}:: 1

End

Program 3.3 (continued)

$H_{ealthy}L_{ife}S_{tyle}$ = [0, 0.5]. Then, the computational process may be continued, with the outcome:

$$eval_{kf_{new\,case}}\underbrace{\left(((0.67, 0.67)(1, 1)), \cdots, ((0, 0.5)(1, 0.87))\right)}_{\substack{attribute`s\ values\ ranges\ once\ normalized\ and \\ respective\ QoI\ and\ DoC\ values}} :: 1 :: 0.82 \qquad (11)$$

Now, the *new case* may be portrayed on the *Cartesian* plane in terms of its *QoI* and *DoC*, and by using clustering methods [23] it is feasible to identify the cluster (s) that intermingle with the *new one*. The *new case* is compared with every *retrieved case* from the cluster using a similarity function *sim*, given in terms of the average of the modulus of the arithmetic difference between the arguments of each case of the selected cluster and those of the *new case*. Thus, one may have:

$$\begin{array}{l} retrieved_{case_1}\left(\left(((0.70, 0.70)(1, 1)), \cdots, ((0.5, 0.75)(1, 0.97))\right)\right) \\ \qquad :: 1 :: 0.83 \\ \qquad \vdots \\ retrieved_{case_{63}}\underbrace{\left(((0.63, 0.63)(1, 1)), \cdots, ((0.33, 0.5)(1, 0.98))\right)}_{normalized\ cases\ that\ make\ the\ retrieved\ cluster} \\ \qquad :: 1 :: 0.84 \end{array} \qquad (12)$$

Table 3.1 The coincidence matrix for *CB* model

Target	Predictive	
	True (1)	False (0)
True (1)	95	11
False (0)	18	149

Assuming that every attribute has equal weight, for the sake of presentation, the *dis(imilarity)* between new_{case} and the $retrieved_{case1}$, i.e., $new_{case \longrightarrow 1}$, may be computed as follows:

$$dis^{DoC}_{new\,case \rightarrow 1} = \frac{\|1-1\| + \cdots + \|0.87 - 0.97\|}{7} = 0.09 \quad (13)$$

Thus, the *sim(ilarity)* for $sim^{DoC}_{new\,case \rightarrow 1}$ is set as *1 – 0.09 = 0.91*. Regarding *QoI* the procedure is similar, returning $sim^{QoI}_{new\,case \rightarrow 1} = 1$. Thus, one may have:

$$sim^{QoI,\,DoC}_{new\,case \rightarrow 1} = 1 \times 0.91 = 0.91 \quad (14)$$

i.e., the product of two measurements is a new type of measurement. For instance, multiplying the lengths of the two sides of a rectangle gives its area, which is the subject of dimensional analysis. In this work the mentioned outcome gives the overall similarity between the new case and the retrieved ones. These procedures should be applied to the remaining cases of the retrieved clusters in order to obtain the most similar ones, which may stand for the possible solutions to the problem. This approach allows users to define the most appropriate similarity methods to address the problem (i.e., it gives the user the possibility to narrow the number of selected cases with the increase of the similarity threshold).

The proposed model was tested on a real data set with 273 examples. Thus, the dataset was divided in exclusive subsets through the ten-folds cross validation [24]. In the implementation of the respective dividing procedures, ten executions were performed for each one of them. Table 3.1 presents the coincidence matrix of the *CB* model, where the values presented denote the average of 25 (twenty five) experiments. A perusal to Table 3.1 shows that the model accuracy was 89.4% (i.e., 244 instances correctly classified in 273). Thus, the predictions made by the *CB* model are satisfactory, attaining accuracies close to 90%. The sensitivity and specificity of the model were 89.6 and 89.2%, while *Positive* and *Negative Predictive Values* were 84.1 and 93.1%, denoting that the model exhibits a good performance in the evaluation of kidney function.

3.6 Conclusion

The main contribution of this work is a monitoring system for kidney assessment. Its distinguishing features are a balance between the number of necessary inputs and monitoring performance, being mobile-friendly, and featuring an online learning component that enables the automatic recalculation and evolution of the monitoring process upon the addition of new cases. Indeed, this article focus on a new approach to problem solving in the area of *Personal Assistants*, where through the use of *CBR* it is conceivable to classify new unlabeled cases based on a repository of past ones. Being built not only on quantitative and qualitative information or knowledge, and taking into consideration unknown, incomplete, or self-contradictory information or knowledge, provides an unprecedented base for personal healthcare care assistants. It also provides a learning doorway for those who assist persons with physical incapacities, enabling them not only to live self-sufficiently in the society, but also providing a means of measuring its effectiveness, and even creating tailored stuff. These results make it easier to perform because of the use of RapidMiner, which adapts not only the entire *ETL* (*E*xtract, *T*ransform, and *L*oad) process, but also provides procedures for data mining and machine learning, including preprocessing and data visualization, predictive analysis and statistics, and unfolding.

Future work will not only consider the development of predictive models to allow the patient to obtain an estimate of the evolution of their health status, but also analyze the complexity and the high cost of building the *CBs* that make difficult, if not impossible, to evaluate a *CBR* system, especially a knowledge-intensive one.

Acknowledgements This work has been supported by COMPETE: POCI-01-0145-FEDER-007043 and FCT—Fundação para a Ciência e Tecnologia within the Project Scope: UID/CEC/00319/2013.

References

1. Levey, A.S., Bosch, J.P., Lewis, J.B., Greene, T., Rogers, N., Roth, D.: A more accurate method to estimate glomerular filtration rate from serum creatinine: a new prediction equation. Ann. Internal Med. **130**, 461–470 (1999)
2. Shemesh, O., Golbetz, H., Kriss, J.P., Myers, B.D.: Limitations of creatinine as a filtration marker in glomerulopathic patients. Kidney Int. **28**, 830–838 (1985)
3. Perrone, R.D., Madias, N.E., Levey, A.S.: Serum creatinine as an index of renal function: new insights into old concepts. Clin. Chem. **38**, 1933–1953 (1992)
4. Aamodt, A., Plaza, E.: Case-based reasoning: foundational issues, methodological variations, and system approaches. AI Commun. **7**, 39–59 (1994)
5. Richter, M.M., Weber, R.O.: Case-Based Reasoning: A Textbook. Springer, Berlin (2013)
6. Neves, J.: A logic interpreter to handle time and negation in logic databases. In: Muller, R., Pottmyer, J. (eds.) Proceedings of the 1984 Annual Conference of the ACM on the 5th Generation Challenge, pp. 50–54. Association for Computing Machinery, New York (1984)
7. Neves, J., Machado, J., Analide, C., Abelha, A., Brito, L.: The halt condition in genetic programming. In: Neves, J., Santos, M.F., Machado, J. (eds.) Progress in Artificial Intelligence. LNAI, vol. 4874, pp. 160–169. Springer, Berlin (2007)
8. Kakas, A., Kowalski, R., Toni, F.: The role of abduction in logic programming. In: Gabbay, D., Hogger, C., Robinson, I. (eds.) Handbook of Logic in Artificial Intelligence and Logic Programming, vol. 5, pp. 235–324. Oxford University Press, Oxford (1998)

9. Pereira, L., Anh, H.: Evolution prospection. In: Nakamatsu, K. (ed.) Studies in Computational Intelligence, vol. 199, pp. 51–64. Springer, Berlin (2009)
10. Machado, J., Abelha, A., Novais, P., Neves, J., Neves, J.: Quality of service in healthcare units. In: Bertelle, C., Ayesh, A. (eds.) Proceedings of the ESM 2008, pp. 291–298. Eurosis—ETI Publication, Ghent (2008)
11. Fernandes, A., Vicente, H., Figueiredo, M., Neves, M., Neves, J.: An adaptive and evolutionary model to assess the organizational efficiency in training corporations. In: Dang, T.K., Wagner, R., Küng, J., Thoai, N., Takizawa, M., Neuhold, E. (eds.) Future Data and Security Engineering. Lecture Notes on Computer Science, vol. 10018, pp. 415–428. Springer International Publishing, Cham (2016)
12. Fernandes, F., Vicente, H., Abelha, A., Machado, J., Novais, P., Neves J.: Artificial neural networks in diabetes control. In: Proceedings of the 2015 Science and Information Conference (SAI 2015), pp. 362–370, IEEE Edition (2015)
13. Quintas, A., Vicente, H., Novais, P., Abelha, A., Santos, M.F., Machado, J., Neves, J.: A case based approach to assess waiting time prediction at an intensive care unity. In: Arezes, P. (ed.) Advances in Safety Management and Human Factors. Advances in Intelligent Systems and Computing, vol. 491, pp. 29–39. Springer International Publishing, Cham (2016)
14. Silva, A., Vicente, H., Abelha, A., Santos, M.F., Machado, J., Neves, J., Neves, J.: Length of stay in intensive care units—A case base evaluation. In: Fujita, H., Papadopoulos, G.A. (eds.) New Trends in Software Methodologies, Tools and Techniques, Frontiers in Artificial Intelligence and Applications, vol. 286, pp. 191–202. IOS Press, Amsterdam (2016)
15. Vicente, H., Dias, S., Fernandes, A., Abelha, A., Machado, J., Neves, J.: prediction of the quality of public water supply using artificial neural networks. J. Water Supply: Res. Technol. —AQUA **61**, 446–459 (2012)
16. Mendes, R., Kennedy, J., Neves J.: Watch thy neighbor or how the swarm can learn from its environment. In: Proceedings of the 2003 IEEE Swarm Intelligence Symposium (SIS 2003), pp. 88–94. IEEE Edition (2003)
17. Blanco, X., Rodríguez, S., Corchado, J.M., Zato, C.: Case-based reasoning applied to medical diagnosis and treatment. In: Omatu, S., Neves, J., Corchado, J.M., Santana, J., Gonzalez, S. (eds.) Distributed Computing and Artificial Intelligence. Advances in Intelligent Systems and Computing, vol. 217, pp. 137–146. Springer, Berlin (2013)
18. Begum, S., Ahmed, M.U., Funk, P., Xiong, N., Folke, M.: Case-based reasoning systems in the health sciences: a survey of recent trends and developments. IEEE Trans. Syst. Man Cybern. Part C (Applications and Reviews) **41**, 421–434 (2011)
19. Oliveira, T., Silva, A., Neves, J., Novais, P.: Decision support provided by a temporally oriented health care assistant. J. Med. Syst. **41**(1), paper 13 (2017)
20. Hofmann, M., Klinkenberg, R.: RapidMiner: Data Mining Use Cases and Business Analytics Applications. CRC Press, Boca Raton (2013)
21. Zhao, H., Doshi, P.: Towards automated restful web service composition. In: Proceedings of the 2009 International Conference on Web Services, pp 189–196. IEEE Edition (2009)
22. Carneiro, D., Novais, P., Andrade, F., Zeleznikow, J., Neves, J.: Using Case-based reasoning and principled negotiation to provide decision support for dispute resolution. Knowl. Inf. Syst. **36**, 789–826 (2013)
23. Figueiredo, M., Esteves, L., Neves, J., Vicente, H.: A data mining approach to study the impact of the methodology followed in chemistry lab classes on the weight attributed by the students to the lab work on learning and motivation. Chem. Educ. Res. Pract. **17**, 156–171 (2016)
24. Haykin, S.: Neural Networks and Learning Machines. Pearson Education, New Jersey (2009)

Part III
Health

Chapter 4
Visual Working Memory Training of the Elderly in VIRTRAEL Personalized Assistant

Miguel J. Hornos, Sandra Rute-Pérez, Carlos Rodríguez-Domínguez, María Luisa Rodríguez-Almendros, María José Rodríguez-Fórtiz and Alfonso Caracuel

Abstract Personal assistants using emerging technologies are showing a great potential to provide an important impact in different aspects of daily human life currently and during the near future. They are usually intended to help people with especial needs. The one presented in this chapter, called VIRTRAEL, is especially designed to assess, stimulate and train several cognitive skills that experience a decline as people age, reason why its target public is the elderly, as well as the therapists who treat them. VIRTRAEL is made up of different types of exercises, each of them specifically designed to evaluate and stimulate a different cognitive function. After presenting an overview of our tool, we focus on one of its exercises, the one devoted to the classification and memorization of images, which is intended to train the visual working memory. We also present a configuration tool that allows the therapists to customize and adapt each exercise to the preferences and needs of a given user. Moreover, we show some results of a pilot study carried out with a sample of elderly people.

4.1 Introduction

The use of emerging technologies to develop personal assistants to help elderly people to improve their quality of life at diverse ambits of daily life is a duty of our society nowadays. This will contribute not only to help the elderly to lead a more

---

M.J. Hornos (✉) · C. Rodríguez-Domínguez · M.L. Rodríguez-Almendros · M.J. Rodríguez-Fórtiz
Software Engineering Department, ETSIIT, CITIC, University of Granada, Granada, Spain
e-mail: mhornos@ugr.es

S. Rute-Pérez · A. Caracuel
CIMCYC, University of Granada, Granada, Spain

A. Costa et al. (eds.), *Personal Assistants: Emerging Computational Technologies*, Intelligent Systems Reference Library 132, DOI 10.1007/978-3-319-62530-0_4

independent and satisfying life but also to save a significant amount of economic costs (in health, mobility, social inclusion, etc.) to governments and institutions.

One of these ambits of application would be the one related to cognitive functions. As is known, many older people usually suffer some degree of decline in their cognitive skills, which involves different symptoms, such as a reduction in problem solving capacity, less ability to reason and to maintain the attention focus, and forgetfulness, to mention only a few. Several studies [12, 30, 34] have demonstrated that an adequate cognitive stimulation helps to decrease the rate of both age-related intellectual decay and cognitive decline.

Moreover, emerging technologies can help to provide new training opportunities developed as exercises and tests implemented as serious games [4, 11, 18, 24, 37] to delay the decline in certain cognitive, sensory-motor, social, and even emotional functions that people suffer as they age.

One of the cognitive abilities that can be trained using technologies is the classification and memorization of images, task that aims to train the Visual Working Memory (VWM) ability. This cognitive skill is responsible for the processing and maintenance of the visual information necessary to solve problems in a short time [23]. It is mainly supported by the Visuospatial Sketchpad, a component of the most accepted Working Memory Model [1]. The Visuospatial Sketchpad maintains the visual objects and their spatial position for a short time. This maintenance of the visual information is really important because allows the Working Memory to operate independently of the direct stimulation of the retina [19]. Daily, we use this ability to face many of our activities, such as mental calculations to check if we have enough money when buying at the supermarket, information retaining that we perceive through the sight to understand it when reading a book, or even interpretation and remembering of traffic signs once they have passed when driving.

VWM is one of the main cognitive functions in the performance of activities of daily living. From the small amount of information processed in the VWM we can represent in a stable way and understand the visuospatial environment and the world [19]. This ability allows us to perform complex activities that require keeping irrelevant information out of our thoughts and actions [13]. Processing information in the VWM allows us to plan and guide our behaviour in a wide variety of situations [19], such as tracking traffic signals when driving, the colour and shape of the medication we just took [22], or our movement in a room when the light suddenly goes out [15].

VWM deteriorates with aging, dementia or Alzheimer's disease. A VWM dysfunction may be a key factor in the widespread deterioration of other cognitive functions that depend on it [13]. A specific deterioration of the VWM has been seen in asymptomatic individuals but with early-onset Alzheimer's disease [22]. This deficit implies the loss of the capacity to represent objects as an integrated whole, and consequently the impaired performance of daily activities. However, cognitive interventions produce changes in the cerebral activation of healthy elderly people and with dementia, provoking a neural plasticity that facilitates learning [36]. Daily intellectual training with tasks such as puzzle solving has beneficial effects on the working memory of older people. These effects are greater when the training environment demands a higher concentration [20]. Also, computerized cognitive training improves visual memory, even in people at high risk of cognitive impairment [5].

In this paper, we firstly present an overview of a web-based system that we have developed, called VIRTRAEL (VIRtual TRAining for the ELderly). It has been specially designed to both assess and stimulate the elderly users' cognitive skills by means of a series of exercises and tests implemented as serious games. Some of these exercises are implemented as 2D serious games [17, 28, 29] and other ones as 3D serious games [27]. All of them encourage a greater user engagement and motivation and each of them are devoted to train at least one of these cognitive functions: memory, attention, planning and reasoning. Users, not only the elderly but also their therapists and carers, can access to VIRTRAEL platform using a personal computer, a laptop or a tablet.

Elderly people have difficulties to use technologies, and more specifically elders with cognitive impairments. For this reason, VIRTRAEL has been designed taking into account several usability guidelines oriented to elderly people, including the one corresponding to W3C.[1] Besides, its usability has been tested by a group of users in a pilot study [29]. As VIRTRAEL can be executed in a computer or in a mobile device, such as a tablet, the users interact with it using a mouse (in the first case) or their fingers on the touch screen (in the second case). Older people usually prefer this last form of interaction because it is more direct and intuitive and makes them easier the use of the application.

VIRTRAEL, as a personal assistant, guides the users through the cognitive training process, evaluates user responses, provides feedback, and makes decisions to adapt the difficulty levels of the exercises and personalize them. Adaptation rules are defined by the therapists so that an intelligent system uses them to fit the exercises to each user at run-time.

Among all the exercises included in VIRTRAEL, we focus on the one dedicated to train the classification and memorization of images. In the design and implementation of the exercises, we have taken into account all the theoretical aspects described by the psychologists integrated in the development team.

The rest of the chapter is organized as follows: Sect. 4.2 revises some related tools, analysing their main characteristics and comparing our tool with respect to them. Section 4.3 gives an overview of VIRTRAEL, describing it and presenting some of the exercises or games that it includes. Section 4.4 presents the exercise devoted to classification and memorization of images, as well as some results of a

[1]https://www.w3.org/WAI/older-users/developing.html.

pilot study that we have carried out with a sample of elderly people performing such exercise, and a configuration tool that allows personalizing the exercise. Finally, Sect. 4.5 outlines the conclusions and future work.

4.2 Related Work

Currently, there is an increase in the number of computer-based applications that are marketed as sound scientific tools to improve the cognitive functions of older adults, but there are few systematic studies [35]. VIRTRAEL has been compared to four widely used programs for cognitive stimulation: REHACOM, COGMED, COGNIFIT, and GRADIOR. There are some similarities between all programs, such as a certain degree of customization and the reporting of user performance and progress. However, there are specific differences between VIRTRAEL and these programs about visual memory training. Below we summarize some of the features of the programs and of the differences found.

REHACOM.[2] The program makes a log of the execution of the user, provides feedback and automatically adjusts the level of difficulty to the user progress. It allows the therapist to perform the individualization of exercises through images for each user and the adjustment of parameters such as training and reaction time, or the type of encouragement and feedback. It includes training in visual memory along with other cognitive components, but it does not include the teaching of any memorization strategy, such as the categorization one. REHACOM was developed for rehabilitation after brain injury due to stroke, multiple sclerosis or trauma [6]. We only found one study [16] with REHACOM in the elderly (30 healthy people aged between 65 and 80). Half of the sample was allocated to a REHACOM training group while the other half was allocated to a physical training group. Results showed significant improvements in both groups but there were no differences with them. Consequently, studies with larger samples are needed.

COGMED.[3] Monitoring and feedback of this tool have to be done by a certified Coach. There are adjusted difficulties levels based on the user performance but the exercises cannot be personalized. The program is focused on the improvement of the working memory and includes training in VWM. It does not allow the use of categorization as a learning and memory strategy. Hyer et al. [10] performed the first study to use COGMED with older adults with MCI. In this study, the participants improved in working memory but they did not report transfer effects for tasks dissimilar enough from the trained tasks. A review [33] showed a lack of research about COGMED with people older than 60. Just one study included people older

[2]http://www.rehacom.us/index.php/aboutrehacom.

[3]http://www.cogmed.com/.

than 50 years, but also children younger than 10 years. Therefore, there is a limitation for extracting sound conclusions.

COGNIFIT.[4] Therapists can customize sessions but not exercises. This tool makes a record of the results of the exercises and adjusts the training according to the user performance. It includes VWM tasks but no training in memorization strategies. Eighteen older participants showed verbal (non-visual) WM improvement after an online training with COGNIFIT [8]. Studies with larger samples are also needed.

GRADIOR.[5] Professionals can design interventions personalized and adapted to the level of performance and the needs of each person. It includes training tasks in visual attention but no strategies are taught or trained for improving VWM. To the best of our knowledge, there is a lack of sound scientific evidences about GRADIOR effectiveness.

The main advantage of VIRTRAEL compared to the programs analyzed is the inclusion of training in categorization as a memorization strategy. The other programs analyzed simply use massive presentation of images and users are not taught to use any strategy to learn or remember them.

We have also found other two important general differences. One of them is that only VIRTRAEL includes a character (selected by the user from a set of characters) that will be in charge of providing the instructions and feedback to the user. The other important difference is that VIRTRAEL is open-source and free.

4.3 VIRTRAEL Description

VIRTRAEL[6] (VIRtual TRAining for the ELderly) is a web-based system specially intended for cognitive assessment and stimulation of elderly people. The idea is that its end users (i.e. older people, considered as patients) can carry out a series of exercises from anywhere with Internet connection to train certain cognitive skills with the aim of improving them or at least maintaining them. At the same time, VIRTRAEL allows a therapist to configure and supervise the activities carried out by his/her patients within the platform. In addition, it allows the communication among people using the application, and is prepared to provide adaptation and collaboration mechanisms [26]. To carry out all this, VIRTRAEL includes three key tools:

Communication tool. The idea is to use this tool to avoid or at least decrease the sense of isolation that older people often have, by encouraging the communication and even the collaboration with other users. To do this, it provides a forum to exchange messages among people involved in the cognitive evaluation and

[4]https://www.cognifit.com.

[5]http://www.ides.es/gradior.

[6]http://www.everyware.es/webs/virtrael/.

stimulation process, as well as a chat to allow real-time collaboration between end users and their therapists. As a result, we will get the user engagement with the application.

Configuration tool. This tool is intended to be used by either the administrator or the therapists to assign elderly users (patients) to the corresponding therapist, as well as carers to a given elder user. It also allows the therapists to supervise which exercises are being carried out by their patients and consult what results they are getting at real time. Taking into account this information, the therapists can choose which exercises are more adequate to be carried out by each of their patients at a given moment, and their best execution order, as well as configure them more accurately. Thus, for example, a therapist who is monitoring one of his/her patients can decide, depending on the user's degree of success in the exercise that is carrying out, whether the next exercise planned for such user should be performed or skipped. Moreover, this tool can be considered as an *authoring tool*, because it allows creating personalized exercises based on the one included in VIRTRAEL. Hence, a therapist can customize a given exercise for a specific user using this tool. This last aspect of this tool will be addressed in Sect. 4.2.

Cognitive evaluation and stimulation tool. It can be considered the main tool of VIRTRAEL, and includes 18 different exercises distributed in 15 predefined work sessions. The first two sessions and the last two ones are respectively devoted to assess the cognitive skills of the user before and after the stimulation sessions, which are the rest ones, where the mentioned exercises are carried out to train such cognitive abilities. Each exercise has been designed to either evaluate or stimulate (depending on the session where it is executed) one or more of the following cognitive functions: memory, attention, reasoning and planning. As a general rule, a character presents the appropriate instructions to the user at the beginning of each exercise, telling him/her what must do in it. After reading them, the user should act accordingly to complete the exercise. Thus, for example, the user must select one or more elements from a set of (textual or graphical) items in some cases, while s/he must write the corresponding element in other cases, to mention only a few of the possible actions to be carried out in the different exercises. The following figures show screenshots of some of the exercises included in this tool: the one presented in Fig. 4.1 is intended to stimulate attention, the one in Fig. 4.2 corresponds to an exercise to train reasoning, the ones in Fig. 4.3 are related to an exercise devoted to work planning, while the ones in Fig. 4.4 correspond to an exercise designed to mainly stimulate visuospatial attention, and more specifically its modalities of divided, alternating and selective attention, but also to secondarily train categorical reasoning.

The character used in VIRTRAEL behaves as an intelligent virtual assistant [21] that acts as a mediator between the user and the application. It interacts with the user through a textual interface. Sound has not been added because the incompatibility and the lack of support for some of the more common web browsers. In our case, the "intelligence" of this character is shown in two ways: (1) orienting the user to repeat an exercise or trial again if his/her results are not good, and (2) adapting its response when congratulating or communicating bad results to the

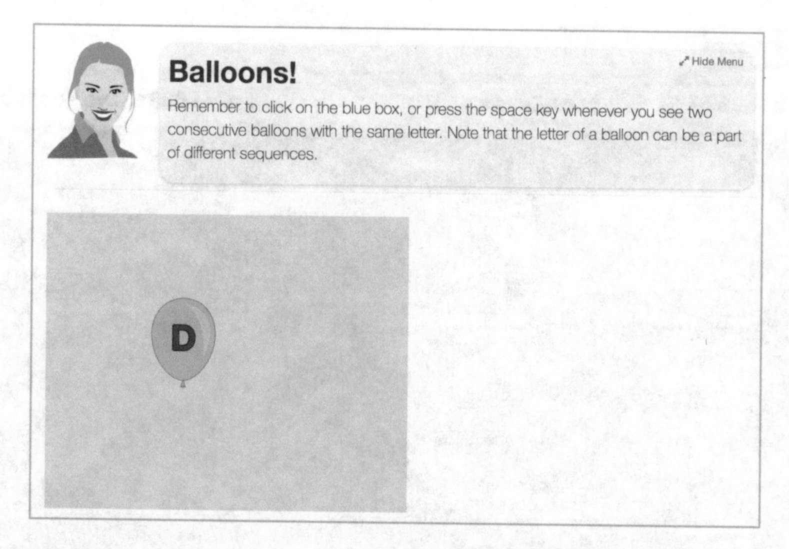

Fig. 4.1 Exercise on balloons in movement, especially designed to train attention

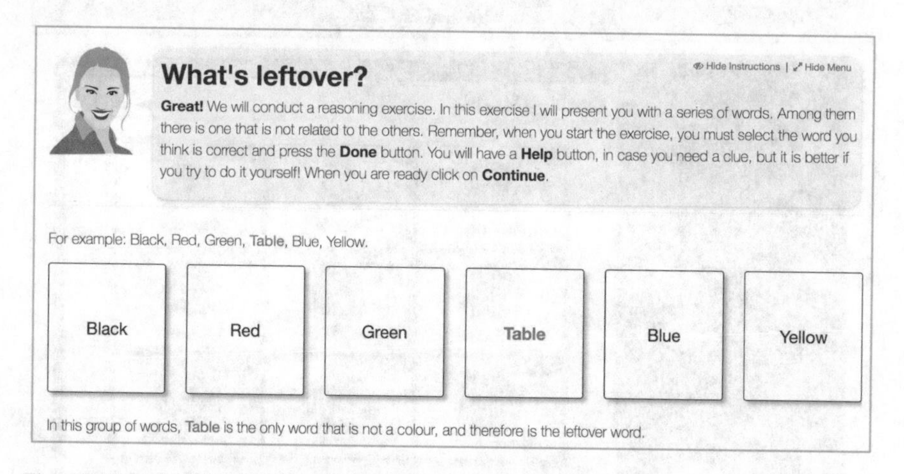

Fig. 4.2 Exercise on semantic series, designed to train reasoning

user, encouraging him/her to do it better next time. The users who participated in the VIRTRAEL pilot study confirm us that they felt accompanied and encouraged by this character, because they had someone to pay attention to and be accountable to, as indicated by Borini et al. [2]. Moreover, we have checked that the existence of this character to interact with is a way to facilitate older people to focus their attention on different key moments, such as realizing the beginning of a new exercise or understanding instructions.

While the user is performing an exercise, the values of several variables (such as playtime, number of failures, hits and omissions, etc.) are registered to evaluate

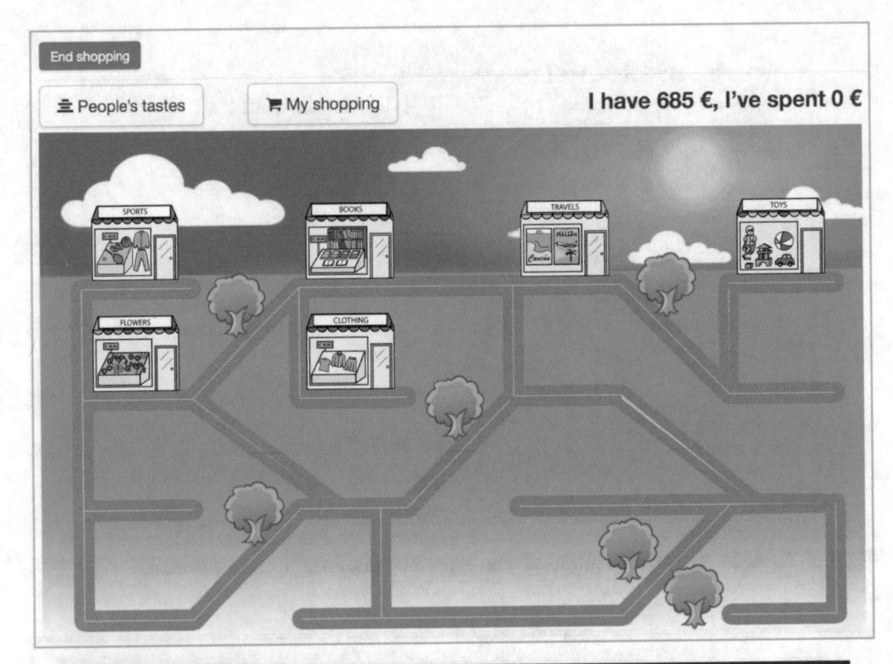

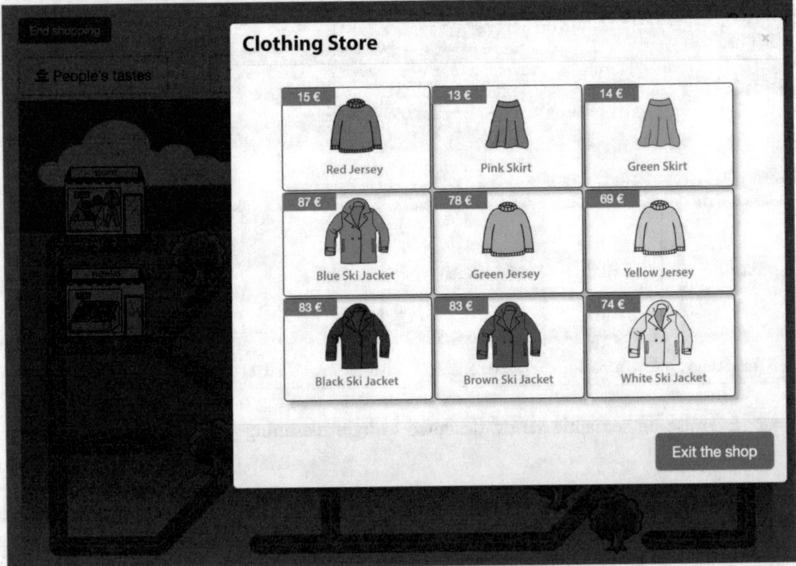

Fig. 4.3 Exercise on purchase of several gifts with a certain budget, designed to train planning: The upper screenshot shows the different shops, while the bottom one shows the items available in the shop selected by the user

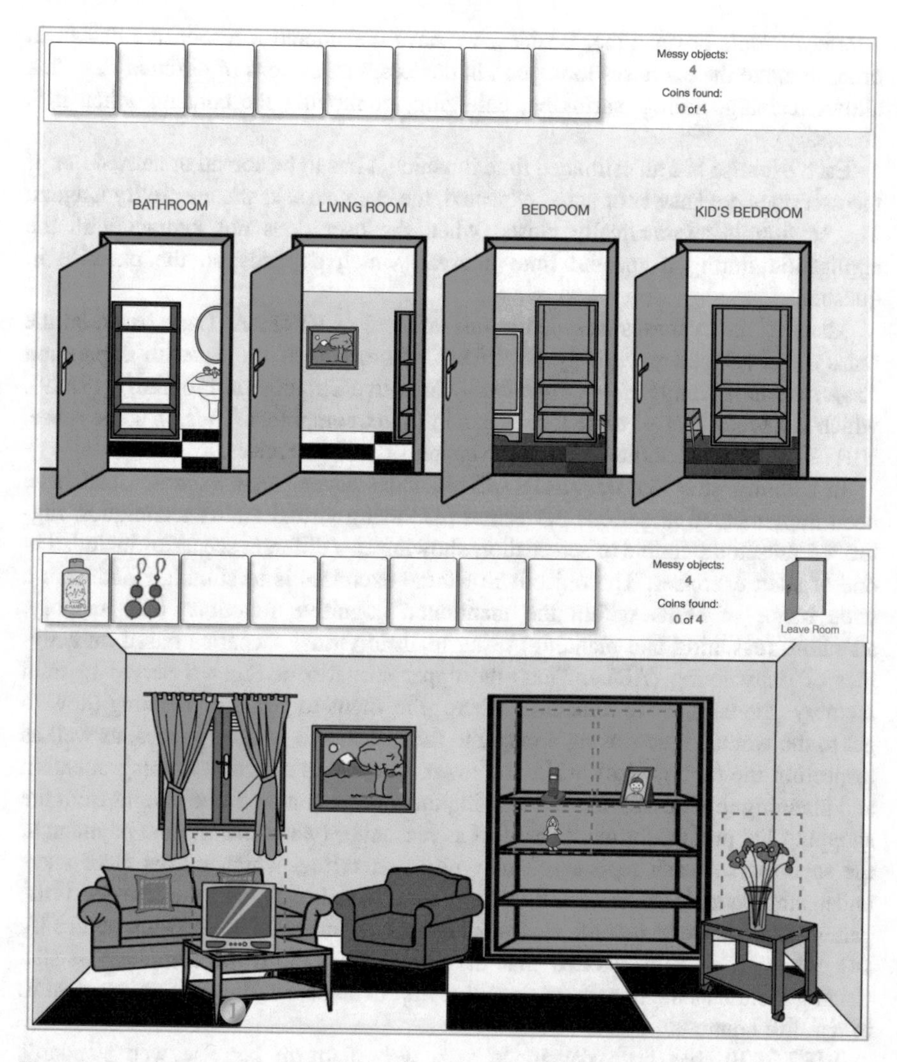

Fig. 4.4 Exercise to put messy objects in the right place and collect the coins lying on the floor, designed to train visuospatial attention and categorical reasoning: The upper screenshot shows the corridor that gives access to the different rooms of the house, while the bottom one presents one of these rooms (the living room in this case) once the user has entered in it, showing the items that can be moved surrounded by a rectangle with dashed line

his/her degree of success. In addition, these values are taken into account to adapt not only the difficulty level of the exercise but also its user interface at run time, i.e. while the user is carrying out the exercise. In VIRTRAEL architecture, there is a separation between the model and the view. In fact, there are several views and interface components per each exercise. They are used depending on the device and web browser in which the corresponding exercise is being executed, so that the user interface adapts at run time. Besides, we have considered a responsive design in order to make the exercises look good in devices with screens of different size (this allows resizing, hiding, shrinking, enlarging, or moving the contents when it is necessary).

Each exercise has an estimated time in which it has to be solved or carried out. If the user does not answer or act as expected, the exercise asks the user to try it again. The session is automatically closed when the user does not interact with the application during a specific time interval, which depends on the exercise in question.

Many of the exercises included in this last tool of VIRTRAEL are based on the ones we developed previously for PESCO (acronym of the Spanish expression *Programa de EStimulación COgnitiva* – Cognitive Stimulation Program) [17, 29], which is a tool intended to be locally run in Linux computers. Thus, in some sense, VIRTRAEL provides the web-based version of such exercises.

In addition, this tool of VIRTRAEL includes several new exercises that have been implemented as serious 3D games [27] using virtual reality techniques. Figure 4.5 presents a couple of screenshots showing two different scenarios included in one of such exercises. The objective of these exercises is to stimulate and train a wide range of skills within the mentioned cognitive functions (i.e. memory, attention, reasoning and planning) using realistic virtual scenarios based on activities of daily living (ADL). Thus, the upper scenario in Fig. 4.5 serves to train memory (the user has to remember where s/he wants to go), and planning (how to get to the wanted place, using a map and the orientation in open spaces, as well as respecting the traffic rules), while the lower scenario is aimed at training attention and planning (by looking for and locating the objects that are to be bought from the shopping list previously made) as well as reasoning (when making the payment at the supermarket cash register). This type of exercises, which makes elder users understand better what to do and feel more motivated, facilitates the transfer of the gaming experience to real life similar situations, as indicated in previous studies [3, 14], where it is demonstrated that the ecological validity of computer-assisted training facilitates the transfer of cognitive improvements to the performing of ADL in real life contexts.

VIRTRAEL has been developed as a web platform because web standards provide *portability*. Thus, anyone with a device that can run a web browser and from anywhere with connexion to Internet can access and use our platform. This technology also allows simultaneously creating and running different instances of one or more exercises. And this makes possible that a series of users can use the services provided by VIRTRAEL at the same time, from different geographical locations and with a much lower (temporary and economic) cost than if the same exercises had to be performed in the therapist's consultation. All of it thanks to patients can perform them at their own home or whatever other place with access to Internet where they are. This avoids the temporary and/or economic costs of having to go to the consultation, the time in the waiting room of the consultation, etc.

Fig. 4.5 Two scenarios of a 3D game included in VIRTRAEL: The upper one corresponds to a street, showing a map in the right upper corner, while the lower one presents shelves with items inside a supermarket

In addition, VIRTRAEL allows the therapists to be more efficient performing their work and thus to be able to attend to a greater number of patients, irrespective of where these reside.

Moreover, as other of our aims is providing a program of exercises that can be performed by elderly people without the constant support of a therapist, a series of *proactive adaptations* have been incorporated into the platform. Consequently, these adaptations have to be carried out at run-time, and they should be done in a twofold sense, because it is necessary to adapt: (1) the user interface, due to the diversity of features that the different devices used by the users to access to the platform can have; and (2) the contents and structure of the exercises, according to the level of difficulty selected and the skills shown by the user while performing the exercise.

To carry out both types of adaptation, we have designed an *on-the-fly decision-making mechanism* which has been internally implemented by means of a *rule-based system* [9], consisting of a series of "if <predicate>, then <action/s>" rules. Each of them is triggered when its corresponding predicate is satisfied. A predicate can make reference to the cognitive skills, personal information, daily habits and hobbies (or any other information considered of interest in the future) stored in the user profile, as well as to the results of partially performing an exercise, the type of exercise, the number of trials performed, the kind of device used by the user, etc. Some examples of actions to be executed when the predicate of a rule is satisfied could be: modifying the value assigned to one or more variables of the exercise (e.g. the total possible choices as answer to a question, the amount of on-screen elements or the number of steps) in order to increase or decrease its level of difficulty and/or change its workflow, as well as adjusting some of its presentation features (e.g. interaction mode, fonts, colours, and/or widgets to be used). The execution of the action/s associated to a rule implies the update of the exercise instance that the user is performing to include the new properties in it. Thus, for instance, if a user is not able to complete an exercise after several attempts, then a simplified exercise (e.g. with fewer elements) and/or a tutorial on how to carry out the exercise could be shown to the user just after the execution of the corresponding rule.

Before finish this section, we should mention that two previous studies with a representative sample of elderly users have been carried out [29]. In the first study participated forty-three elderly people (65% women) with a mean age of 74 years (SD = 10.9). Just 35% of them had previously used a computer. As some results showed that there was room for usability improvements, several changes had been made motivated by such results. Others results showed that the tasks included in the cognitive test module demonstrated its concurrent validity with traditional cognitive evaluation tests. In the second study, seventy older people were allocated in an experimental and a control group for determinate efficacy of the cognitive training with VIRTRAEL. Results showed that the stimulation exercises were effective for improving attention, verbal working memory and planning skills in the elderly.

4.4 Classification and Memorization of Images Exercise

The Classification and Memorization of Images exercise does not use a massed presentation of stimuli as a memory improvement strategy. Unlike other computer-based exercises, what is asked to older people is not to memorize the images but rather to classify them into categories. Categorization is our ability to group stimuli into categories that indicate semantic relationships between stimuli [7]. This ability is innate and very adaptive, since for example it helps us to detect and avoid potentially dangerous objects and situations [32]. This skill can be improved throughout life but from the age of 60 there is a gradual decrease in performance in object categorization, especially in incongruent contexts. This decrease could impair the recognition of objects in daily visual tasks and environments [25]. Training with classification tasks has shown effectiveness in improving visual memory. For example, training in categorization strategies for memorization tasks improved visual memory, as opposed to movement discrimination strategies [31].

This exercise also trains object discrimination. The target images are displayed together with distractors that belong or not to the same category of the image that users have previously visualized. Older people should decide and indicate for each stimulus (target or distractor) whether or not it was among those who were shown in the initial phase. In order to have an adequate performance in this recognition task, the visual information of the previously categorized objects must be kept active in the VWM [7].

The exercise has two phases. The first one shows four categories of objects and asks the user to select images of objects from a set of images in the screen and to classify them accordingly (see Fig. 4.6). Some examples of categories are fruits, meals, parts of the body, sports, tools, transports or professions. Categories and objects are randomly selected in each execution to present different elements belonging to a diverse range of domains, thus preventing exercise from being boring.

The user has to select an object from above and move it to a cell into a category from below. The proposal of this exercise is showing the users that the objects can be classified in order to help their mind to memorize them easily. Besides, this provides more hints during the recovering process from the memory, being a very useful strategy of memorization.

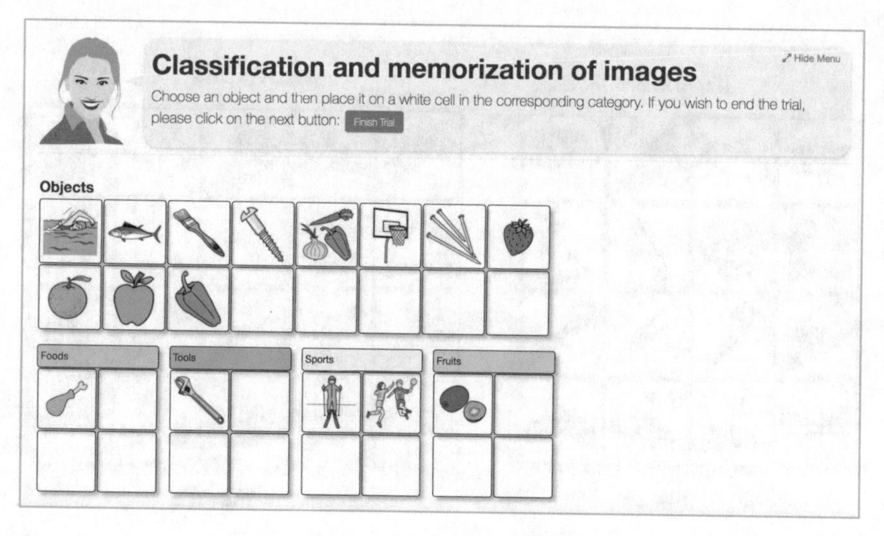

Fig. 4.6 Classification phase of the Classification and Memorization of Images exercise

When the user has finished the exercise, VIRTRAEL registers the time used in performing it, and the number of fails, successes and omissions (i.e. objects without classification). If these values are optimal, the user is considered as “prepared” and can continue with the second phase of the exercise. In other case, the first phase of the exercise should be repeated again. The psychologists integrated in our development team know the optimal values and VIRTRAEL has been programmed according to them. Thus, for example, in order to progress to a higher level, the face-to-face general rule of traditional cognitive training has been applied. In fact, to reach the next level of difficulty, it is necessary to achieve at least 80% of the total score at the current level.

There is a character acting as an assistant that describes the exercise and explains step by step what the user has to do and how to interact with the application to solve the exercises. The assistant also informs the user about his/her successes and fails to provide him/her a feedback, and encourage the user to continue.

The second phase of the exercise consists of a maximum of eight trials to train memory. Each one shows images of objects that have to be memorized during a specific time. The number of images and time in each trial vary taking into account the performance of the user and his/her results (see Fig. 4.7). As before, the psychologists integrated in our development team have decided when and how each trial takes into account the previous user interaction results. As shown in Fig. 4.7, there are differences between trials according the previous hits and fails of the user. Thus, the left screenshot in such figure, which corresponds to trial 7 out of a total of 8 (as indicated in the text at the bottom), presents an array of 16 images to be memorized in 60 s (the capture was done in the ninth second of them, as indicated in the counter that is shown to the user above the array), while the right screenshot, corresponding to the last test (8/8), presents a matrix of 20 images to be memorized in 75 s.

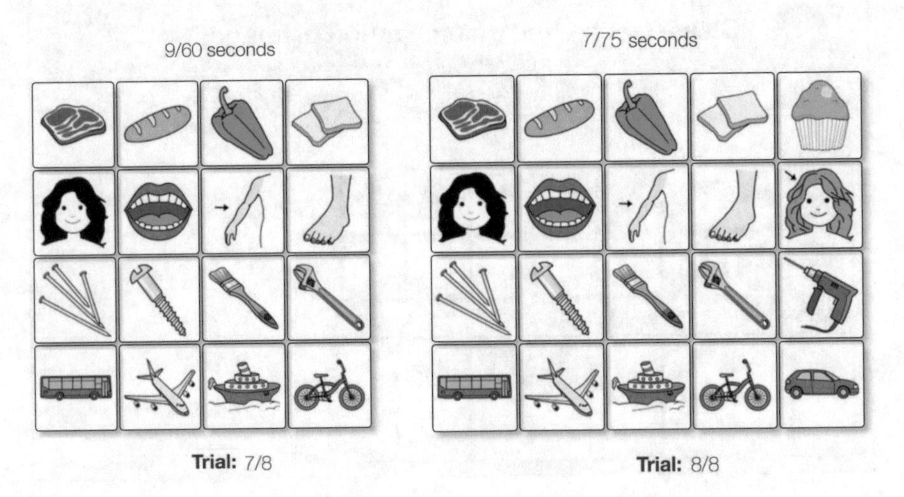

Fig. 4.7 Memorization phase of the Classification and Memorization of Images exercise

After memorizing the objects in each trial, the user has to select only the objects that have been memorized by him/her from a set of images shown in the screen (see Fig. 4.8). If the user applies the classification strategy previously mentioned to memorize them, it will be easier.

The assistant guides the user through the trials, and also provides him/her feedback just after the execution of each trial, by informing the user his/her results, as shown in Fig. 4.9.

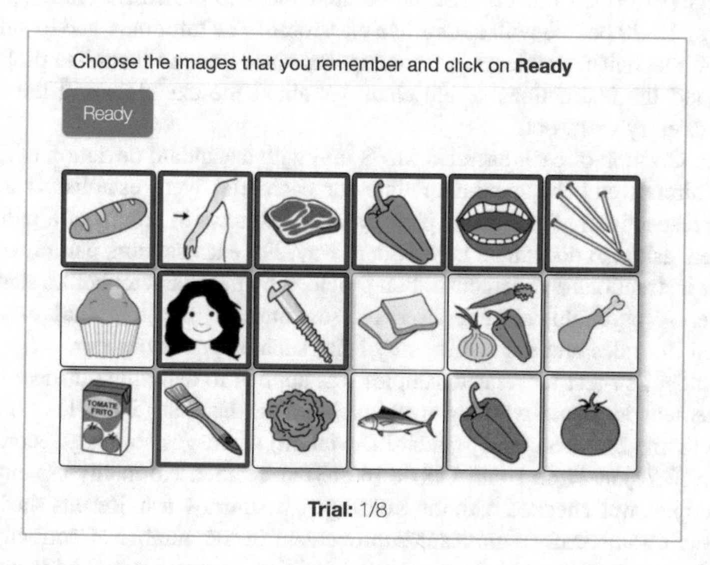

Fig. 4.8 Images selected by the user (highlighted with a thicker border in blue)

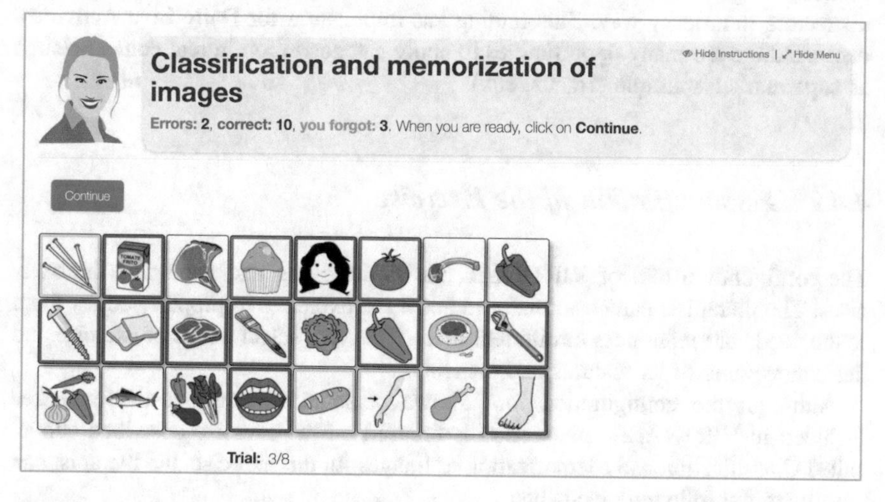

Fig. 4.9 Feedback on the performance of each trial provided by the assistant

4.4.1 Pilot Study

A pilot study has been conducted to determine the effectiveness of the Classification and Memorization of Images exercise for improving the visual working memory in the elderly. Participants without dementia were recruited from community computer centers. One hundred and seventy nine people over 65 years performed the exercise three times, with an interval of 2 weeks between each training session. Twenty of the users performed the exercise alone at home and the rest were divided into groups of 15-20 people with a tutor in a classroom. The tutor only had to help some users with technical problems in the computer, since none of them had problems to understand the instructions of the character about the exercises and the steps to follow to carry them out.

Mean duration of each exercise was 8 min with a standard deviation of 2.9 min. The minimum and the maximum time for each trial were established at 1 and 30 min respectively. Participants performed the exercise in groups of 8 people, but they were asked to do it in an independent way. For each session, participants must read the instructions and demonstration provided by the character before starting the exercise. A psychologist supervised all sessions for checking that participants followed the rules and did not use any help, such as pen and paper.

A Student's t-test for related samples was applied to determine if there was any improvement in visual working memory between the first (Time 1) and the last session (Time 2). Mean (and Standard Deviation) of images correctly recorded was 109,64 (60,77) at Time 1 and 128,48 (65,53) at Time 2. Normality assumption of both samples was checked with the Kolmogorov-Smirnov test. Results showed that there was a statistically significant improvement in the number of correctly memorized images between the first and third training sessions [$t = -3.642$; $p < 0.001$]. Our hypothesis for this great improvement was that users learnt that applying the trained categorization strategy allows them to discriminate between target and distractors in an easy way. This finding has implication for Daily Live Activities, because there are many opportunities to apply categorization in real contexts (such as supermarket, multiple errands, etc.).

4.4.2 Personalization of the Exercise

The configuration tool of VIRTRAEL allows designing and managing the exercises. The therapists can customize and adapt the exercises to properly adjust them to the needs or preferences of different users. They can select a type of exercise and determine some of its features and contents.

Although the configuration tool allows personalize some of the exercises included in VIRTRAEL, this section is focused on the exercise described above, called Classification and Memorization of Images. In this exercise, the therapist can customize the following elements:

- Categories management:
  - Add a category: The therapist creates a new category name.
  - Modify a category: The therapist can modify the name of a category.
  - Delete a category: The category is eliminated from the set of categories.
- Objects management (see Fig. 4.10):
  - Add an object to a category: The therapist has to provide an object name and an associated image.
  - Modify an object of a category: The therapist can modify the name and the image associated to the selected object of a category.
  - Delete an object of a category: The category is eliminated from the corresponding category.
- Characteristics of the second phase of the exercise (see Fig. 4.11):
  - Number of trials.
  - Memorization time for each trial.
  - Maximum trial time (to select the memorized objects).
  - Position of the objects in screen. It can be fixed or random.
  - Number of objects of each category that are displayed in each trial.

In order to adapt the exercise to the previous user performance, the therapist can determine a set of rules of adaptation that are used at run-time to change the difficulty level of the exercises. There are rules of adaptation in each phase.

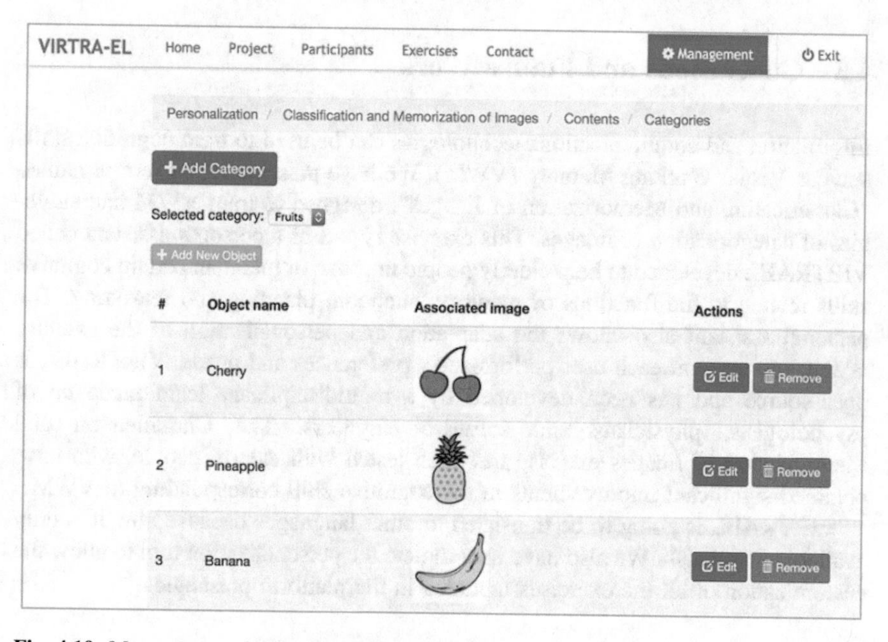

Fig. 4.10 Management of objects in a selected category

VIRTRA-EL Home Project Participants Exercises Contact Management Exit

Personalization / Classification and Memorization of Images / Contents / Trials

+ Add

#	Memory time	Maximum trial time	Number of categories	Number of objects	Position	Actions
1	45 seconds	180 seconds	2	4	Random	Edit Remove
2	30 seconds	180 seconds	2	6	Random	Edit Remove
3	30 seconds	180 seconds	4	4	Fixed	Edit Remove

Fig. 4.11 Edition of exercise characteristics to customize it

- First Phase: It allows determining whether the user can start or not with the second phase depending the required time to complete the exercise, as well as the number of failures, hits and omissions.
- Second Phase: The characteristics of the exercise depend on the performance of the user on the previous trials. If the user carries out one trial correctly, the difficulty level of the exercise is increased, and it is decreased in the other case. The difficulty level is defined by the number of trials, the number of categories and objects in each category for each trial, the memorization time available in each trial and the maximum time to finalize each trial.

4.5 Conclusions and Future Work

Information and communications technologies can be used to train cognitive skills such as Visual Working Memory (VWM). We have presented the exercise named "Classification and Memorization of Images", designed to train VWM and strategies of categorization of images. This exercise is part of a personal assistant called VIRTRAEL developed to help elderly people improve or maintain certain cognitive skills related to the functions of memory, attention, planning and reasoning. The personal assistant also allows the adaptation and personalization of the exercise taking into account each user performance, preferences and needs. VIRTRAEL is open-source and has been developed by a multidisciplinary team made up of psychologists, physicians and software engineers. The Classification and Memorization of Images exercise has been tested with elderly people, who have obtained significant improvements in the cognitive skill corresponding to VWM.

VIRTRAEL is going to be translated to other languages because now it is only available in Spanish. We also have to complete the personalization tool to allow the customization of all the exercises included in the platform presented.

References

1. Baddeley, A., Eysenck, M.W., Anderson, M.C.: *Memory*. Psychology Press (2010)
2. Borini, E., Damiano, R., Lombardo, V., Pizzo, A.: Dramasearch. Character-mediated search in cultural heritage. In: 2nd Conference on Human System Interactions (HSI'09). pp. 554–561. IEEE (2009). http://www.di.unito.it/~rossana/concorso_ROMA/pagina_pubblicazioni/f49-2009-IEEE_HSI.pdf
3. Chevignard, M., Pillon, B., Pradat-Diehl, P., Taillefer, C., Rousseau, S., Le Bras, C., Dubois, B.: An ecological approach to planning dysfunction: script execution. Cortex **36**(5), 649–669 (2000)
4. Cowan, B., Sabri, H., Kapralos, B., Cristancho, S., Moussa, F., Dubrowski, A.: SCETF: Serious game surgical cognitive education and training framework. In: 2011 IEEE International Games Innovation Conference, pp. 130–133. IEEE (2011)
5. Coyle, H., Traynor, V., Solowij, N.: Computerized and virtual reality cognitive training for individuals at high risk of cognitive decline: systematic review of the literature. Am. J. Geriatr. Psychiatry **23**(4), 335–359 (2015)
6. d'Amato, T., Bation, R., Cochet, A., Jalenques, I., Galland, F., Giraud-Baro, E., Pacaud-Troncin, M., Augier-Astolfi, F., Llorca, P.M., Saoud, M., Brunelin, J.: A randomized, controlled trial of computer-assisted cognitive remediation for schizophrenia. Schizophr. Res. **125**(2–3), 284–290 (2011)
7. Freedman, D.J., Assad, J.A.: Neuronal mechanisms of visual categorization: an abstract view on decision making. Ann. Rev. Neurosci. **39**, 129–147 (2016)
8. Gigle, K.L., Blomeke, K., Shatil, E., Weintraub, S., Reber, P.J.: Preliminary evidence for the feasibility of athome online cognitive training with older adults. Gerontechnology **12**(1), 26–35 (2013)
9. Hayes-Roth, F.: Rule-based systems. Commun. ACM **28**(9), 921–932 (1985)
10. Hyer, L., Scott, C., Atkinson, M.M., Mullen, C.M., Lee, A., Johnson, A., Mckenzie, L.C.: Cognitive training program to improve working memory in older adults with MCI. Clin. Gerontol. **39**(5), 410–427 (2016)
11. Ijsselsteijn, W., Nap, H.H., de Kort, Y., Poels, K.: Digital game design for elderly users. In: Proceedings of the 2007 Conference on Future Play, pp. 17–22. ACM (2007)
12. Jean, L., Bergeron, M.È., Thivierge, S., Simard, M.: Cognitive intervention programs for individuals with mild cognitive impairment: systematic review of the literature. Am. J. Geriatr. Psychiatr. **18**(4), 281–296 (2010)
13. Jost, K., Bryck, R.L., Vogel, E.K., Mayr, U.: Are old adults just like low working memory young adults? Filtering efficiency and age differences in visual working memory. Cereb. Cortex **21**(5), 1147–1154 (2011)
14. Krasny-Pacini, A., Limond, J., Evans, J., Hiebel, J., Bendjelida, K., Chevignard, M.: Context-sensitive goal management training for everyday executive dysfunction in children after severe traumatic brain injury. J. Head Trauma Rehabil. **29**(5), E49–E64 (2014)
15. Ko, P.C., Ally, B.A.: Visual cognition in Alzheimer's disease and its functional implications. INTECH Open Access Publisher (2011)
16. Lee, Y.M., Jang, C., Bak, I.H., Yoon, J.S.: Effects of computer-assisted cognitive rehabilitation training on the cognition and static balance of the elderly. J. Phys. Therapy Sci. **25**(11), 1475–1477 (2013)
17. López-Martínez, A., Santiago-Ramajo, S., Caracuel, A., Valls-Serrano, C., Hornos, M.J., Rodríguez-Fórtiz, M.J.: Game of gifts purchase: Computer-based training of executive functions for the elderly. In: IEEE 1st International Conference on Serious Games and Applications for Health (SeGAH), pp. 1–8. IEEE (2011)
18. Melenhorst, A.S.: Adopting communication technology in later life: The decisive role of benefits. Eindhoven University of Technology (2002)
19. Nie, Q.Y., Müller, H.J., Conci, M.: Hierarchical organization in visual working memory: From global ensemble to individual object structure. Cognition **159**, 85–96 (2017)

20. Nozawa, T., Taki, Y., Kanno, A., Akimoto, Y., Ihara, M., Yokoyama, R., Kotozaki, Y., Nouchi, R., Sekiguchi, A., Takeuchi, H., Miyauchi, C.M., Ogawa, T., Goto, T., Sunda, T., Shimizu, T., Tozuka, E., Hirose, S., Nanbu, T., Kawashima, R.: Effects of different types of cognitive training on cognitive function, brain structure, and driving safety in senior daily drivers: a pilot study. Behavioural Neurology, 2015, article ID 525901, 18 p (2015). doi:10.1155/2015/525901
21. Ostinelli, R.: The Composite Intelligence of Virtual Assistants (2007). http://www.uxmatters.com/mt/archives/2007/10/the-composite-intelligence-of-virtual-assistants.php
22. Parra, M.A., Abrahams, S., Logie, R.H., Méndez, L.G., Lopera, F., Della Sala, S.: Visual short-term memory binding deficits in familial Alzheimer's disease. Brain **133**(9), 2702–2713 (2010)
23. Pertzov, Y., Heider, M., Liang, Y., Husain, M.: Effects of healthy ageing on precision and binding of object location in visual short term memory. Psychol. Aging **30**(1), 26–35 (2015)
24. Rego, P., Moreira, P.M., Reis, L.P.: Serious games for rehabilitation: a survey and a classification towards a taxonomy. In: 5th Iberian Conference on Information Systems and Technologies, pp. 1–6. IEEE (2010)
25. Rémy, F., Saint-Aubert, L., Bacon-Macé, N., Vayssière, N., Barbeau, E., Fabre-Thorpe, M.: Object recognition in congruent and incongruent natural scenes: A life-span study. Vis. Res. **91**, 36–44 (2013)
26. Rodríguez-Domínguez, C., Carranza-García, F., Rodríguez-Almendros, M.L., Hurtado-Torres, M.V., Rodríguez-Fórtiz, M.J.: Real time user adaptation and collaboration in web based cognitive stimulation for elderly people. In: 13th International Conference Distributed Computing and Artificial Intelligence (DCAI), pp. 367–375. Springer (2016)
27. Rodríguez-Fórtiz, M.J., Rodríguez-Domínguez, C., Cano, P., Revelles, J., Rodríguez-Almendros, M.L., Hurtado-Torres, M.V., Rute-Pérez, S.: Serious games for the cognitive stimulation of elderly people. In: IEEE 4th International Conference on Serious Games and Applications for Health (SeGAH), pp. 1–7. IEEE (2016)
28. Rute-Pérez, S., Rodríguez-Domínguez, C., Rodríguez-Fórtiz, M.J., Hurtado-Torres, M.V., Caracuel, A.: Training working memory in elderly people with a computer-based tool. In: International Conference on Computers Helping People with Special Needs, pp. 530–536. Springer (2016)
29. Rute-Pérez, S., Santiago-Ramajo, S., Hurtado, M.V., Rodríguez-Fórtiz, M.J., Caracuel, A.: Challenges in software applications for the cognitive evaluation and stimulation of the elderly. J. NeuroEng. Rehabil. **11**(88), 1–10 (2014)
30. Selzer, M., Clarke, S., Cohen, L., Duncan, P., Gage, F.: Textbook of neural repair and rehabilitation, vol. 2. Cambridge University Press, Medical Neurorehabilitation (2006)
31. Sarma, A., Masse, N.Y., Wang, X.J., Freedman, D.J.: Task-specific versus generalized mnemonic representations in parietal and prefrontal cortices. Nat. Neurosci. **19**(1), 143–149 (2016)
32. Schenk, S., Minda, J.P., Lech, R.K., Suchan, B.: Out of sight, out of mind: Categorization learning and normal aging. Neuropsychologia **91**, 222–233 (2016)
33. Shinaver, I., Entwistle, P.C., Söderqvist, S.: Cogmed WM training: reviewing the reviews. Appl. Neuropsychol.: Child **3**(3), 163–172 (2014)
34. Talassi, E., Guerreschi, M., Feriani, M., Fedi, V., Bianchetti, A., Trabucchi, M.: Effectiveness of a cognitive rehabilitation program in mild dementia (MD) and mild cognitive impairment (MCI): a case control study, Archives of Gerontology and Geriatrics, 44(Suppl.), pp. 391–399 (2007)
35. Tusch, E.S., Alperin, B.R., Ryan, E., Holcomb, P.J., Mohammed, A.H., Daffner, K.R.: Changes in neural activity underlying working memory after computerized cognitive training in older adults. Frontiers Aging Neurosci. **8**, 1–14 (2016)
36. van Os, Y., de Vugt, M.E., van Boxtel, M.: Cognitive interventions in older persons: do they change the functioning of the brain? BioMed. Res. Int. article ID 438908, 14 p. (2015). doi:10.1155/2015/438908
37. Wiemeyer, J., Kliem, A.: Serious games in prevention and rehabilitation—a new panacea for elderly people? Eur. Rev. Aging Phys. Activity **9**(1), 41–50 (2011)

Chapter 5
Personal Robot Assistants for Elderly Care: An Overview

Ester Martinez-Martin and Angel P. del Pobil

Abstract The world's population is ageing and, with that, new social issues arise, especially in terms of healthcare and daily activities. Despite the preference for human professional healthcare, the new socio-economic situation and the decrease in care personnel make necessary to give support to the process of caregiving. In this context, Robotics can be considered as a solution since it can provide healthcare support, help in performing daily tasks, and/or increase the feeling of autonomony and self management. This paper is an overview of the existing robotic technologies for elderly care, analysing their benefits for the elderly.

5.1 Introduction

A social analysis reveals a clear tendency to ageing population (see Fig. 5.1). This fact may impact on several socio-economic issues such as the sustainability of families, the economic growth, or the ability of states and communities to provide resources for older citizens [1]. But, probably, one of the most important social deficits is the change in family structure since it leads to leave elderly people, who see their capability to look after themselves weakened, with fewer options for care. As a consequence, most elderly people end in nursing homes that, despite being a good alternative, it is not their desire. Actually, older people prefer to be in their own homes, even if that means living alone [2]. However, some factors coming from ageing like limited mobility, make this independent life become an impossible dream.

In this context, Robotics can play a main role since it may overcome the arisen challenges such as healthcare, assistance in case of emergency, or treatment compliance [3]. Nevertheless, as pointed out by Diaz et al. [4], despite the well-recognized potential benefit of using Robotics in providing physical assistance, the elderly will not use them just because they need them. In fact, attitudes towards robots in elderly

E. Martinez-Martin (✉) · A.P. del Pobil
Universitat Jaume I, Avda. Sos Baynat s/n, Castellón, Spain
e-mail: emartine@uji.es

A.P. del Pobil
e-mail: pobil@uji.es

A. Costa et al. (eds.), *Personal Assistants: Emerging Computational Technologies*, Intelligent Systems Reference Library 132, DOI 10.1007/978-3-319-62530-0_5

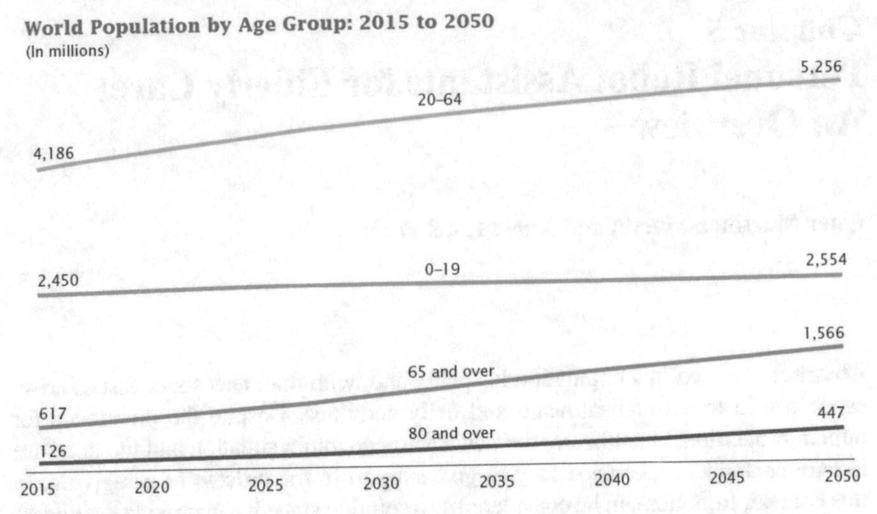

Fig. 5.1 World population by age group: 2015 to 2050 *Source* U.S. Census Bureau, 2013; International data base

care are systematically sceptical. So, the acceptance of this technology depends on the proper understanding of the elders (and their caregivers) expectancies with regard to care and assistance, especially in the case of robotic solutions supporting social behaviour. Therefore, the key to success is to be perceived as compatible with elderly life-style, acceptable by their relatives and caregivers and respectful with the image of how they want their homes to be.

Thus, one of the first issues to be solved is to determine what role these robots play in people's everyday lives. On this regard, one of the early research in this area was presented by Dautenhahn et al. [5]. This user-centred study revealed that the majority of people want the robot to be an assistant, while the role of machine/appliance was the second preferred option. Finally, the less popular choice was having a robot as a servant. Going a step further, Huber et al. [6] identified social roles for a domestic care robot enabling independent living by means of a robot-stimulus-free approach. In this way, an assistive robot could be provided with basic traits while also adapting to the individual needs of its owner over time. From these needs and preferences, one can derive concrete robot behaviours for the most interesting basic behaviours and role adaptation behaviours.

Another important issue is what the system is for. So, two different categories can be defined in terms of assistive robots for elderly [7]:

1. Rehabilitation robots: these robots are focused on physical assistive technology (e.g. artificial limbs [8–10], lifting and walking robots, robotic beds [11, 12], smart wheelchairs [13, 14], active orthoses and exoskeletons [15–19]) and are not primarily communicative (is not meant to be perceived as a social entity)
2. Assistive social robots: these robots can be perceived as social entities that communicate with their users.

This paper takes a walk around different robotic solutions intended to *care for the elderly*, that is, the technology covering from home chores to interactive cognitive training and companionship. In addition, this technology is discussed based on the elderly needs and requirements. Note that, despite its great importance in mobility rehabilitation tasks and the wide research in this area, rehabilitation robots are not covered in this paper since it does not integrate any social skill (even not communication in some cases).

5.2 Assistive Social Robots

Assistive social robotics refers to a research area defined as the intersection of assistive robotics, aimed at aiding human users through interactions with robots, and socially interactive robotics, intended to socially engage human users through interactions with robots [20].

This research can be used to improve the quality of life of people with age-related health issues by being provided with the capabilities necessary to support them in their daily living activities and cognitive tasks. In this context, two different types of robot can be identified:

- Companion robots: aimed to enhance health and psychological well-being of elderly users by providing companionship
- Service robots: designed to perform daily tasks and support basic activities and mobility.

5.2.1 Companion Robots

One of the first efforts and more commonly-used functions of assistive social robots for the elderly focused on robots in the role of a companion. Mainly, a companion robot could be defined as a type of assistant robot which goal is to proactively assist elderly in everyday tasks, reduce stress and depression, enhance social interaction and elicite emotional responses.

In this regard, pet-like robots can help in assisted therapy and activity since they provide psychological, physiological and social effects. Basically, these robots behave actively, physically interacting with their users and stimulating their affection. The most widely used robotic pet is PARO (Personal Assistant RObot) [21], a seal robot developed especially for therapy. In fact, as shown in Fig. 5.2, PARO looks like a baby of harp seal, which has white fur and, beneath it, tactile, vision, audition and posture sensors allow it to perceive the human interaction. This vast set of sensors provides PARO with the capabilities of sensing a user's touch, recognizing a limited amount of speech, expressing a small set of vocal utterances and moving its head and front flippers, giving the illusion that it responds to its environment and to

Fig. 5.2 The robot Seal PARO (2011)

Fig. 5.3 NeCoRo

the interactions with it such as petting or talking. So, this robot has been used at several facilities for the elderly and at pediatric hospitals by resulting in an improvement of their users's mood as well as in an increase of their activity and communication with each other and their caregivers [22, 23]. In addition, a recent study reveals that the caregivers themselves feel positive emotions towards their PARO, highlighting the instrumental and intrinsic value of using PARO in nursing homes [24].

Another well-known pet robot is NeCoRo, a cat-like robot developed by Omron Corporation (see Fig. 5.3). This robot was created for improving person's quality of life. With that purpose, the design of NeCoRo allows it to express emotions (e.g. satisfaction, surprise, anger or hatred) and appropriately respond to individual's emotions and movements. As shown by Nakashima et al. in their study in an elderly-care facility [25], NeCoRo helps the residents improve the communication among the elderly and makes the environment become calmer, easier, gentler and more comfortable.

Fig. 5.4 Unazuki Kabochan (Nodding Kabochan) [29]

As an alternative to zoomorphic robots, robotic dolls can be also used as a companion robots because of their therapeutic gains [26, 27]. For instance, Yamamoto et al. [28] presented Wonder, a wombat-like robot aimed at delivering daily information to an elderly from a distant place such as local government office or welfare institution, as well as monitoring the elderly.

PIP Co., Ltd., Osaka, and WiZ Co., Ltd., Tokyo, developed Unazuki Kabochan (also known as Nodding Kabochan), a cognitive aid robot capable of talking, singing, and slightly moving as a response to its owner's touch and spoken words [29]. This cartoon-like platform, depicted in Fig. 5.4, helps the elderly improve their cognitive functions thanks to its communication skills [30].

On its behalf, Babyloid is a baby-type robot designed for being taken care of an elderly person requiring nursing care [31]. In this case, as depicted in Fig. 5.5, its appearance is especially designed to be perceived as able to do nothing so that the user is provided with a social role: *caring for a "baby"*. In this task, Babyloid helps its *carers* by expressing its emotions, moods and other feelings on its face and voice. An evaluation of the Babyloid experience in a nursing home for elderly revealed a high acceptance among the patients, opening the door to develop Babyloid-centred therapies for helping the elderly attain motivation.

With the same purpose, other robot systems have been developed to overcome the people loneliness and their stress. For example, the H2020 MARIO project is aimed at developing a companion robot built upon the Kompai R&D Robot. Its main goal is to address the difficult challenges of loneliness, isolation and dementia in older people [32]. A study about its acceptability among people with dementia and healthcare staff, revealed a variety of factors that influences its acceptance such as its friendly face, the knowledge of people's likes and dislikes, its utility, and the ability to connect people to their families and friends [33]. These results set the guidelines for

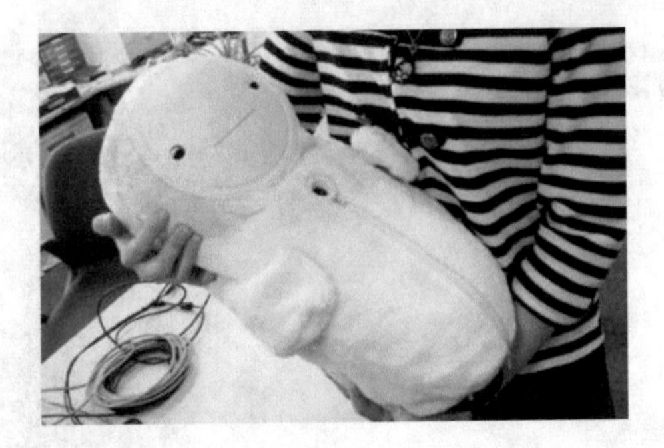

Fig. 5.5 The Babyloid robot

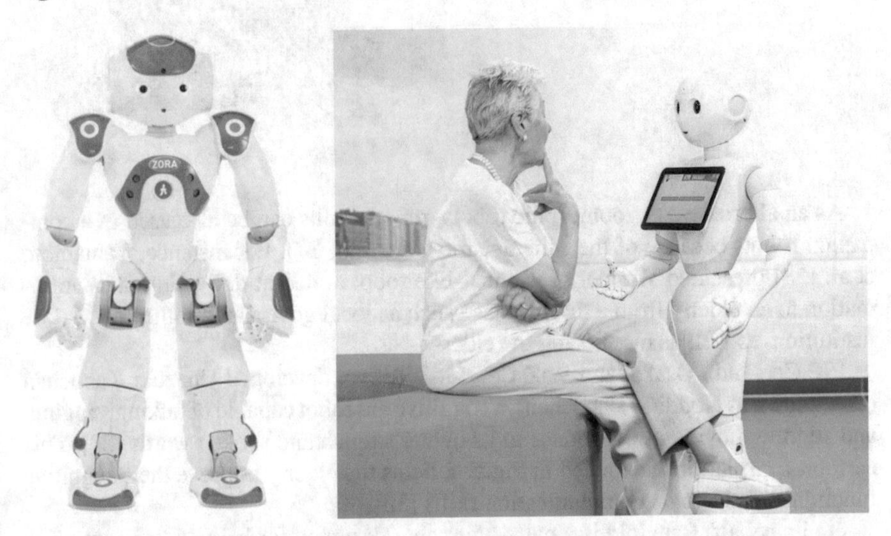

Fig. 5.6 The Zora robot (*left*) and the ZoraBot Pepper (*right*) *Source* http://zorarobotics.be/index.php/en/

technologists to ensure that the MARIO robot integrates the functionalities covering the people's needs and making it more likely to be acceptable.

ZORA (Zorg (Health), Ouderen (Elderly person), Revalidatie (Rehabilitation), Animatie (Animation)) [34] proposed two commercial robotic solutions. On the one hand, it offers the Zora Robot, which is a healthcare application built on a NAO Robot Platform [35] helping elderly with interactive therapeutic and recreational activities (see Fig. 5.6). These humanoid assistants were initially introduced to several Flemish nursing homes on a trial basis, although they will be introduced in U.S. nursing facilities and retirement communities thanks to a recent agreement with ChartaCares.

On the other hand, the Zorabot Pepper is a healthcare application on a Pepper Robot Platform [35]. This robot can welcome patients when they visit a healthcare facility or a hospital, as well as visit them in their rooms to ask how they are and ask for help to nurses if necessary.

Fig. 5.7 BUDDY, a companion robot [36]

Fig. 5.8 AIDO, a home robot

Another commercial solution developed by Blue Frog Robotics is BUDDY [36]. This friendly companion robot is depicted in Fig. 5.7. Mainly, BUDDY increases the care of elderly by providing companionship and assistance in their daily activities by means of, for instance, reminding medication, appointments, deliveries and/or upcoming events. In addition, BUDDY can detect falls and unusual inactivity. A recent study of the use of BUDDY, conducted by the Media Effects Research Laboratory, College of Communications at Penn State, concluded that elderly people welcome the notion of robots when it means physical support, information, entertainment and interactional skills.

In terms of interaction, a similar system is Aido from InGen Dynamics [37] (Fig. 5.8). Although this system has not been designed for elderly people, it can help them improve their lifestyle helping them with the household chores, handling their schedule or keeping their house monitored and safe.

A different point of view is to use entertainment to fight against loneliness. A clear example are the Manzai robots that, as shown in Fig. 5.9, are composed of two robots emulating the two performers of the Manzai, a traditional Japanese comedy. So, firstly, the user enters a keyword so that the system creates and performs a manzai script from web news articles and manzai techniques [38, 39].

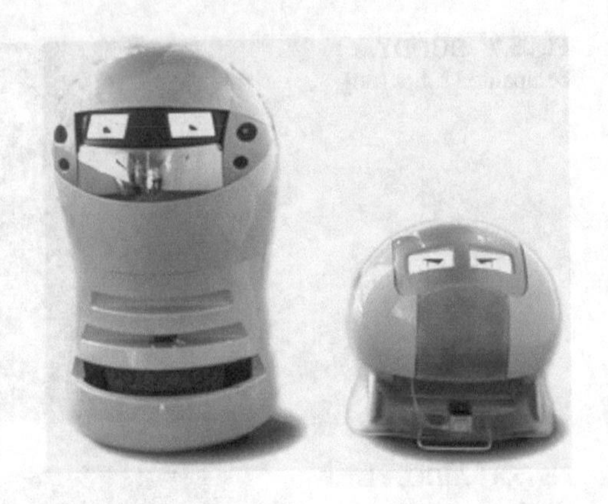

Fig. 5.9 Manzai robots. *Left*: ii-1, Ai-chan. *Right*: ii-2, Gonta *Source* Umetani et al. [39]

5.2.2 *Service Robots*

Service robots can be defined as assistive devices designed to support people living independently by assisting with mobility, completing household tasks, and monitoring health and safety. These are more complex systems because they must adapt to the user's living conditions, what results in being able to safely operate in human-centered environments, interact with people, and assist them in their daily activities.

A high number of service robots has been developed in the last years. For instance, InTouch Health [40] has developed the RP-7 robot, a medical platform remotely operated by a doctor. This mobile platform (Fig. 5.10) can take a patient's pulse, scan vital signs, take pictures and even read case notes. All this data is sent to the *tele*-doctor who can advice medical staff on potentially life-saving actions. However, despite its utility in hospitals, it is not a viable personal assistant for elderly living alone.

Focusing on assistance at home, a mobile robotic assistant is Pearl [41]. Basically, this robot has two main functions:

- Remind elderly people their daily activities such as eating, drinking, taking medicine or using the bathroom. For that, the system is flexible and adaptive, making reminder decision according to the user actions.
- Help elderly people navigate their environments. Thus, Pearl detects people using map differencing: the robot learns a map, and people are detected by significant deviations from the map. This function is particularly important for assistive living facilities.

In the context of smart domestic environments, the RoboCare project started in 2002 [42]. So, its main goal was to build a distributed multi-agent system which provides assistance services for elderly users at home. For that, a combination of software, mobile robots, fixed intelligent sensors and humans is proposed to provide assistance to elderly people were used [43, 44].

Fig. 5.10 The RP-7 robot

With the purpose of providing a system with more functionalities to assist people in their homes, the University of Massachusetts built uBot-5 as a result of its project ASSIST [45]. This small wheeled robot was designed to make user's life easy, to assist them and to notify external caregiver in case of unconsciousness [46].

Another attempt comes from Fraunhofer IPA, who has been developed four generations of Care-O-Bot. The first prototype, Care-O-Bot I was an autonomous mobile platform aimed at communicating with or guiding people. So, its first task was performed in the *Museum für Kommunikation Berlin* where it was in charge of communicating to and interacting with the visitors, autonomously moving around them [47]. Going a step further, the second generation was provided with a manipulator arm, adding the functionality of handling typical objects in a home environment [48]. Later, the third Care-O-Bot generation was launched [49]. In this case, the robot was equipped with a tray to carry objects and safely exchange them with the users. The last prototype, the Care-O-Bot 4, is a friendly, likeable robot with a modular designed that allows it to be configured according to a particular application scenario (e.g. healthcare institutions (Fig. 5.11), hotels or restaurants) [50].

On their behalf, the German research project SERROGA (SERvice RObotics for Gesundheits (Health) Assistance) was aimed at developing robot-based health assistance services for older such that they can remain living in their own homes for as

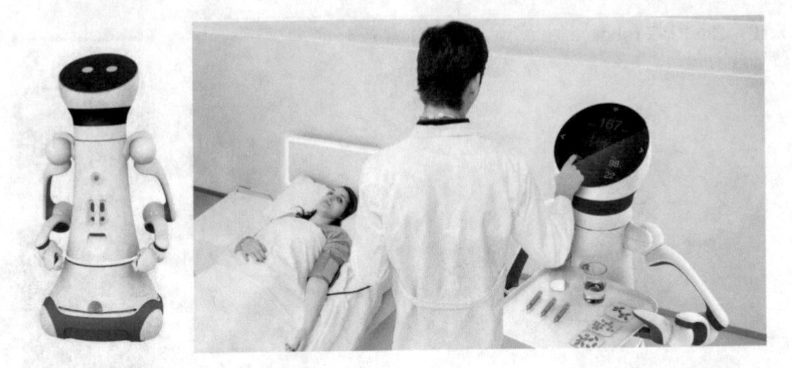

Fig. 5.11 Care-O-Bot *Source* http://www.Care-O-Bot-4.de/

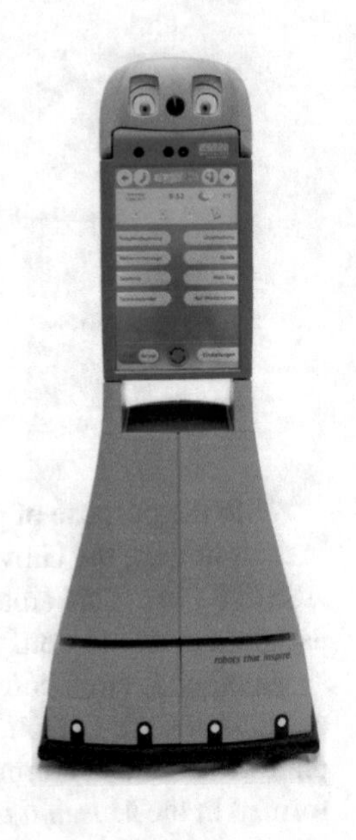

Fig. 5.12 The robot *Max* developed in the project SERROGA *Source* https://www.tu-ilmenau.de/de/neurob/research/robots/tweety-max/

long as possible [51] (Fig. 5.12). As the robot would be cohabitating with the user, an adaptation to the user's needs and preferences was required. As a result, several robot skills (person detection and tracking, fallen person detection, haptic user interface, pulse rate monitoring, exercise monitoring, self-localization, navigation, 3D obstacle detection) as well as behaviours and services (user searching and following, docking and charging, touch-based motion control, dialog manager, GUI, motion exercises) were developed and tested in private senior apartments. The results highlighted that a personal robot assistant has high potential to be accepted by older people as both a useful health assistant and a meaningful social companion.

Similarly, the EU supported project HOBBIT [52] which is intended to develop a socially assistive robot that allows elderly people to feel safe and to stay longer in their homes (Fig. 5.13). Therefore, its main goals are fall prevention measures

Fig. 5.13 The Hobbit robot in three robot tasks: bring an object, pick-up an object from the floor, and detect an emergency situation [52]

(e.g. searching and bringing objects, transporting small items, keeping floors clutter-free, and reminders) as well as emergency detection and handle, while balancing user needs, acceptance and technical performance with a reasonable level of affordability [53]. For that, the concept of *Mutual Care* was introduced such that the robot learns the user habits and preferences in order to adapt its communications and behaviours to them and, at the same time, the user adapts to the robot's intellectual and physical capabilites.

As pointed out by Lammer et al. [54], the use of HOBBIT robot resulted in a high independency of the elderly in care facilities. In addition, a long-term field trial in real private homes of the HOBBIT robot, revealed that users highly appreciated the functions of picking up objects from the floor, transporting objects, emergency recognition, fitness program, and giving reminders [55]. Concerning its usability, despite the intuitive robot handle, the errors in the robot actions led users to frustration. In addition, commands from voice or gesture did not work satisfactorily and processing speed of the whole system was criticized as being too slow. In summary, the HOBBIT usability was negatively influenced by a lack of robustness. Nevertheless, users believed that a market ready version of the robot would be vital for supporting people who are more fragile and more socially isolated.

Along these lines, IBM has recently announced its collaboration with Rice University in the creation of the prototype IBM Multi-purpose Eldercare Robot Assistant (IBM MERA) [56]. The main idea of this research is to combine real-time data generated by ambient sensors with cognitive computing such that the robot could learn about its resident and take better care decisions. In addition, IBM MERA is provided with cameras to read facial expressions and detect falls, sensors to measure elderly heart and breath rates as well as to continuously monitor the resident, and Watson-powered speech recognition ([57]) to know when to call for help.

5.3 Conclusions

World demographic trends reveal an ageing population. This fact results in an increasing need of providing the elderly with the proper care and support in several aspects of their daily life. As a primary goal of elderly is to age in their own homes and the care personnel is limited, assistive technologies are arisen to assist elderly at home, even when they suffer any physical or cognitive disability. In this sense, robots can play a main role since they have the potential to support all the physical, perceptual, and cognitive aspects required in elderly care. However, the key to their success does not only lie in their funcionality, but in their ability to adapt to the user's living conditions and their sociability. As a consequence, several efforts are made to become these robots in a life partner more than just machines.

Along these lines, this paper overviewed state-of-the-art robotic solutions for helping and aiding the elderly in their daily activities such as taking medication, activity scheduling, rehabilitation or asking for assistance in case of emergency. As some studies reported, these developed robotic technologies make easier to live alone for longer time. However, there are still open issues to be overcome, especially in terms of social skills and appearance. Therefore, despite the great advance in the area, there is still a long way to go before leaving our elderly under the care of a robotic system.

Acknowledgements This work has been partially funded by Ministerio de Economía y Competitividad (DPI2015-69041- R), by Generalitat Valenciana (PROMETEOII/2014/028), and by Jaume-I University (P1-1B2014-52).

References

1. Why population aging matters: a global perspective. National Institute on Aging. https://www.nia.nih.gov/research/publication/why-population-aging-matters-global-perspective (2007)
2. Global health and aging. World Health Organization. http://www.who.int/ageing/publications/global_health.pdf (2011)
3. Mitzner, T., Chen, T., Kemp, C., Rogers, W.: Identifying the potential for robotics to assist older adults in different living environments. Int. J. Soc. Robotics (2013)
4. Diaz, M., Saez-Pons, J., Heerink, M., Angulo, C.: Emotional factors in robot-based assistive services for elderly at home. In: 22nd IEEE International Symposium on Robot and Human Interactive Communication (RO-MAN), pp. 711–716. Gyeongju, Korea (2013)
5. Dautenhahn, K., Woods, S., Kaouri, C., Walters, M., Koay, K., Werry, I.: What is a robot companion-friend, assistant or butler? In: IEEE/RSJ International Conference on Intelligent Robots and Systems (IROS), pp. 1192–1197. Alberta, Canada (2005)
6. Huber, A., Lammer, L., Weiss, A., Vincze, M.: Designing adaptive roles for socially assistive robots: a new method to reduce technological determinism and role stereotypes. J. Hum. Robot Interact. **3**, 100–115 (2014)
7. Broekens, J., Heerink, M., Rosendal, H.: Assistive social robots in elderly care: a review. Gerontechnology **8**, 94–103 (2009)
8. Bhattacharyya, S., Konar, A., Tibarewala, D.: A differential evolution based energy trajectory planner for artificial limb control using motor imagery EEG signal. Biomed. Signal Process. Control **11**, 107–113 (2014). doi:10.1016/j.bspc.2014.03.001

9. Juhnke, D.L., Beck, J.P., Jeyapalina, S., Aschoff, H.H.: Fifteen years of experience with integral-leg-prosthesis: cohort study of artificial limb attachment system. J. Rehabil. Res. Dev. **52**(4), 407–420 (2015). doi:10.1682/jrrd.2014.11.0280
10. Sinha, R., van den Heuvel, W.J., Arokiasamy, P.: Adjustments to amputation and an artificial limb in lower limb amputees. Prosthet. Orthot Int. **38**(2), 115–121 (2014). doi:10.1177/0309364613489332
11. Choi, H., Park, J.O., Ko, S.Y., Park, S.: Deflection analysis of a robotic bed on the applied loads and its postures for a heavy-ion therapeutic system. In: 9th International Conference on Robotic, Vision, Signal Processing and Power Applications, pp. 343–350. Springer (2016). doi:10.1007/978-981-10-1721-6_37
12. Wang, C., Savkin, A.V., Clout, R., Nguyen, H.T.: An intelligent robotic hospital bed for safe transportation of critical neurosurgery patients along crowded hospital corridors. IEEE Trans. Neural Syst. Rehabil. Eng. **23**(5), 744–754 (2015). doi:10.1109/tnsre.2014.2347377
13. Gomi, T., Griffith, A.: Developing intelligent wheelchairs for the handicapped, vol. 1458, pp. 150–178. Springer, Berlin, Germany (1998)
14. Shiomi, M., Iio, T., Kamei, K., Sharma, C., Hagita, N.: Effectiveness of social behaviors for autonomous wheelchair robot to support elderly people in japan. PLoS ONE **10** (2015)
15. Anam, K., Al-Jumaily, A.: Active exoskeleton control systems: state of the art. In: International Symposium on Robotics and Intelligent Sensors (IRIS), pp. 988–994 (2012)
16. Benson, I., Hart, K., Tussler, D., van Middendorp, J.J.: Lower-limb exoskeletons for individuals with chronic spinal cord injury: findings from a feasibility study. Clin. Rehabil. **30**(1), 73–84 (2015). doi:10.1177/0269215515575166
17. Dollar, A., Herr, H.: Lower extremity exoskeletons and active orthoses: challenges and state-of-the-art. IEEE Trans. Robotics **24**, 144–158 (2008)
18. Kazerooni, H.: Exoskeletons for human power augmentation. In: IEEE/RSJ International Conference on Intelligent Robots and Systems (IROS). Alberta, Canada (2005)
19. Yan, T., Cempini, M., Oddo, C., Vitiello, N.: Review of assistive strategies in powered lower-limb orthoses and exoskeletons. Robotics Auton. Syst. **64**, 120–136 (2015)
20. Feil-Seifer, D., Mataric, M.: Socially assistive robotics. In: 9th International Conference on Rehabilitation Robotics, 2005. ICORR 2005. Institute of Electrical and Electronics Engineers (IEEE). doi:10.1109/icorr.2005.1501143
21. Wada, K., Shibata, T., Saito, T., Tanie, K.: Effects of robot assisted activity to elderly people who stay at a health service facility for the aged. In: 3 IEEE/RSJ International Conference on Intelligent Robots and Systems (IROS), pp. 2847–2852. Las Vegas, Nevada (2003)
22. Sabanovic, S., Bennett, C., Chang, W.L., Huber, L.: Paro robot affects diverse interaction modalities in group sensory therapy for older adults with dementia. In: IEEE International Conference on Rehabilitation Robotics. Seattle, USA (2013)
23. Wada, K., Shibata, T.: Social and physiological influences of living with seal robots in an elderly care house for two months. In: The 6th International Conference of the International Society for Gerontechnology. Pisa, Italy (2008)
24. Niemela, M., Ylikauppila, M., Talja, H.: Long-term use of paro the therapy robot seal the caregiver perspective. In: 10th World Conference Of Gerontechnology 2016 (ISG 2016). Nice, France (2016)
25. Nakashima, T., Fukutome, G., Ishii, N.: Healing effects of pet robots at an elderly-care facility. In: 9th IEEE/ACIS International Conference on Computer and Information Science, pp. 407–412 (2010)
26. Mitchell, G., McCormack, B., McCance, T.: Therapeutic use of dolls for people living with dementia: a critical review of the literature. Dementia **15**, 976–1001 (2016)
27. Turner, F., Shepherd, M.: Doll therapy in dementia care: a review of current literature. Cumminicare **1** (2014)
28. Yamamoto, H., Miyazaki, H., Tsuzuki, T., Kojima, Y.: A spoken dialogue robot, named wonder, to aid senior citizens who living alone with communication. J. Robotics Mechatron. **14**, 54–59 (2002)

29. Nodding kabochan: cognitive skill aid robot. http://www.aarpinternational.org/resource-library/resources/nodding-kabochan-cognitive-skill-aid-robot (2012)
30. Tanaka, M., Ishii, A., Yamano, E., Ogikubo, H., Okazaki, M., Kamimura, K., Konishi, Y., Emoto, S., Watanable, Y.: Effect of a human-type communication robot on cognitive function in elderly women living alone. Med. Sci. Monit. **18**, CR550–557 (2012)
31. Furuta, Y., Kanoh, M., Shimizu, T., Shimizu, M., Nakamura, T.: Subjective evaluation of use of babyloid for doll therapy. In: IEEE World Congress on Computational Intelligence. Brisbane, Australia (2012)
32. Mario project. http://www.mario-project.eu/portal/ (2015–2018)
33. Casey, D., Felzmann, H., Pegman, G., Kouroupetroglou, C., Murphy, K., Koumpis, A., Whelan, S.: What people with dementia want: designing mario an acceptable robot companion. In: 15th International Conference on Computers Helping People with Special Needs. Linz, Austria (2016)
34. Zorabots. http://zorarobotics.be/index.php/en/ (2017)
35. Softbank robotics. https://www.ald.softbankrobotics.com/en (2017)
36. Buddy, the companion robot. http://www.bluefrogrobotics.com/en/buddy/ (2017)
37. Aido, the next generation home robot. https://www.startengine.com/startup/aido (2016)
38. Umetani, T., Aoki, S., Akiyama, K., Mashimo, R., Kitamura, T., Nadamoto, A.: Scalable component-based manzai robots as automated funny content generators. J. Robotics Mechatron. **28**, 862–869 (2016)
39. Umetani, T., Mashimo, R., Nadamoto, A., Kitamura, T., Nakayama, H.: Manzai robots: entertainment robots based on auto-created manzai scripts from web news articles. J. Robotics Mechatron. **26**, 662–664 (2014)
40. Intouch health. http://www.intouchhealth.com/ (2017)
41. Pollack, M., Engberg, S., Matthews, J., Thrun, S., Brown, L., Colbry, D., Orosz, C., Peintner, B., Ramakrishnan, S., Dunbar-Jacob, J., McCarthy, C., Montemerlo, M., Pineau, J., Roy, N.: Pearl: a mobile robotic assistant for the elderly. In: AAAI Workshop on Automation as Eldercare (2002)
42. Robocare project. http://robocare.istc.cnr.it/ (2002)
43. Bahadori, S., Cesta, A., Grisetti, G., Iocchi, L., Leone, R., Nardi, D., Oddi, A., Pecora, F., Rasconi, R.: Robocare: an integrated robotic system for the domestic care of the elderly. In: Proceedings of Workshop on Ambient Intelligence AI*IA-03. Pisa, Italy (2003)
44. Cesta, A., Cortellessa, G., Pecora, F., Rasconi, R.: Supporting interaction in the robo care intelligent assistive environment. In: Association for the Advancement of Artificial Intelligence (AAAI) Symposium (2007)
45. Amherst, U.: ubot-5. http://www-robotics.cs.umass.edu/index.php/Robots/UBot-5
46. Robot developed by computer scientists to assist with elder care. https://www.umass.edu/newsoffice/article/robot-developed-computer-scientists-assist-elder-care (2008)
47. Graf, B.: Dependability of mobile robots in direct interaction with humans, pp. 223–239. Springer, Berlin, Heidelberg (2005)
48. Graf, B., Hans, M., Schraft, R.: Care-o-bot ii-development of a next generation robotic home assistant. Auton. Robots **16**, 193–205 (2004)
49. Graf, B., Reiser, U., Hagele, M., Mauz, K., Klein, P.: Robotic home assistant care-o-bot 3-product vision and innovation platform. In: IEEE Workshop on Advanced Robotics and its Social Impacts, pp. 139–144. Tokyo, Japan (2009)
50. Ludtke, M.: The service robot care-o-bot 4. CAN Newsletter, pp. 36–39 (2016)
51. Gross, H., Mueller, S., Schroeter, C., Volkhardt, M., Scheidig, A., Debes, K., Richter, K., Doering, N.: Robot companion for domestic health assistance: implementation, test and case study under everyday conditions in private apartments. In: 2015 IEEE/RSJ International Conference on Intelligent Robots and Systems (IROS), pp. 5992–5999 (2015)
52. Hobbit—the mutual care robot. http://hobbit.acin.tuwien.ac.at/index.html (2011)
53. Vincze, M., Zagler, W., Lammer, L., Weiss, A., Huber, A., Fischinger, D., Kortner, T., Schmid, A., Gisinger, C.: Towards a robot for supporting older people to stay longer inde-pendent at home. In: ISR/Robotik 2014; 41st International Symposium on Robotics, pp. 1–7 (2014)

54. Lammer, L., Huber, A., Weiss, A., Vincze, M.: Mutual care: how older adults react when they should help their care robot. In: 3rd International Symposium on New Frontiers in Human-Robot Interaction (2014)
55. Pripfl, J., Kortner, T., Batko-Klein, D., Hebesberger, D., Weninger, M., Gisinger, C., Frennert, S., Eftring, H., Antona, M., Adami, I., Weiss, A., Bajones, M., Vincze, M.: Results of a real world trial with a mobile social service robot for older adults. In: 2016 11th ACM/IEEE International Conference on Human-Robot Interaction (HRI), pp. 497–498 (2016)
56. Cognitive machines assist independent living as a we age. https://www.ibm.com/blogs/research/2016/12/cognitive-assist (2016)
57. Watson speech to text (2017)

Part IV
Personalization

Chapter 6
Personalized Visual Recognition via Wearables: A First Step Toward Personal Perception Enhancement

Hosub Lee, Cameron Upright, Steven Eliuk and Alfred Kobsa

Abstract During the last few years, deep learning has led to an astonishing advancement in visual recognition. Computers now reach *near-human* accuracy in visually recognizing characters, physical objects and human faces. This will certainly allow us to build more intelligent personal assistants that can help users better understand their surrounding environments. However, most visual recognition systems have been designed for user-independent recognition (e.g., Google reverse image search), and not for an individual user. We believe this practice is restricting the technology from helping people who have individual needs. For example, a person with memory problems may want to have a computer that accurately recognizes a few close friends, rather than hundreds of celebrities. To address this issue, we propose a novel wearable system that enables users to create their own visual recognition system with minimal effort. A client running on Google Glass collects images of objects a user is interested in, and sends them to the server with a request for a specific machine learning task: training or classification. The server performs deep learning according to the request and returns the result to Glass. Regarding the training task, our system not only aims to build deep learning models with user generated image data, but also to update the models whenever new data is added by the user. Experiments show that our system is able to train the custom deep learning models in an efficient manner, in terms of the required amount of computing power and training data. Based on the customized deep learning model, the system classifies an image into one of 10 different user-defined categories with 97% accuracy.

Keywords Personalization · Person identification · Object recognition · Google glass · Deep convolutional neural networks · Transfer learning · Finetuning

---

H. Lee (✉) · A. Kobsa
University of California, Irvine, USA
e-mail: hosubl@uci.edu

C. Upright · S. Eliuk
Samsung Research America, Mountain View, USA

A. Costa et al. (eds.), *Personal Assistants: Emerging Computational Technologies*, Intelligent Systems Reference Library 132, DOI 10.1007/978-3-319-62530-0_6

6.1 Introduction

With recent technological advances in wearable computing, users are more likely to collect information about their surroundings. For instance, users could directly capture images of objects through a smart glass rather than a smartphone. Google Glass is a representative wearable that can translate this scenario into reality. Since the camera functionality of Google Glass is always ready to be activated instantaneously by the user's command, it is reasonable to assume that more image data representing users' personal interests could be collected. Consequently, there are more chances to build a user-tailored visual recognition system by utilizing those collected images as training data.

Recently, deep learning has been making a breakthrough in diverse computer vision and pattern recognition problems [12, 16]. Deep learning is a machine learning technique that attempts to extract high-level concepts from data via a complex model composed of hierarchical processing units. The trained deep learning model then utilizes the extracted concepts in making predictions on new data. Deep convolutional neural networks (CNNs), a commonly used type of network, have been widely used in the computer vision community [25]. CNNs are biologically-inspired variants of artificial neural networks, which mimic how the human brain perceives images. These networks consist of multiple layers of filters which hierarchically process segments of the input image. Pooling layers are often added to reduce dimensionality and add translational invariance. Finally, multiple fully connected layers may be used to combine the spatial features and produce a final classification. Specifically, outputs of convolutions in the lower layers are used to represent the primitive element that forms the image (e.g., edge). Then, these representations are integrated in the higher layers to express more abstract concepts (e.g., shape) of the image. With this architecture, we can train the whole network through the standard backpropagation algorithm by using the labeled images as training data. Recent studies proved that CNN-based image classifiers have reached near-human levels on diverse visual recognition tasks [1, 11, 17, 19].

These technological advances can potentially benefit people. However, most deep learning applications thus far have been designed and developed for the general population, not for an individual who has a special need. Imagine that there is a professor who gives a lecture to 300 students. This professor might want to wear Google Glass displaying the names of students during the lecture because it is difficult to memorize all their names. People with visual impairments may want to have Google Glass proatively inform them about the presence of nearby friends. Those with cognitive impairment (e.g., dementia) could use a Google Glass capable of recognizing their personal belongings. This would help when they have memory problems. To realize all of these scenarios, each individual user needs to have a custom machine learning model trained on her/his own image data (i.e., a personalized machine learning model), and utilize the model for recognizing an input image. In this regard, we have defined two main requirements for constructing personalized machine learning models. First, a user should be able to easily collect images with appropriate labels in everyday life. Second, the system should be able to not only train a machine learning model with user generated image data, but update the model whenever the user provides the system with new classes of data.

With these requirements in mind, we developed a novel wearable system that enables users to create their own CNNs without any difficulties. To begin with, our system adopts the first generation of Google Glass for collecting image data representing users' personal interests. By using our Google Glass application named DeepEye, a user can take images of an object of interest and apply whatever label they want. DeepEye then transmits these labeled images to a GPU-equipped Linux server to train the CNN. In general, training deep learning models like a CNN from scratch uses a considerable amount of time on modern GPUs and requires very large volumes of training data. To make the training process more practical, we devised a new training mechanism named chained finetuning. This mechanism was designed to train a new CNN by utilizing the previously trained CNN as a starting point. Experimental results show that chained finetuning allows us to train the CNN while requiring less computational power and training data, compared to conventional training approaches. Most importantly, chained finetuning is effective for expanding the expressive power (i.e., the number of classifiable categories) of the CNN whenever the user collects images of new object classes. DeepEye can also run as an image classifier: if the user asks it to recognize an object from an image, it will show a classification result produced by the server. The server thereby utilizes the CNN trained specifically for the user. Our system showed about 97% accuracy in classifying an image taken by Google Glass into one of 10 user-defined categories.

In summary, the contributions of our work to the field of personal assistants are the following:

- We built a novel wearable system which lets users create their own deep learning-based visual recognition systems without any expertise;
- We proposed a simple, but efficient mechanism for training personalized deep learning models with user-generated image data (chained finetuning); and
- We showed the feasibility of the proposed system including chained finetuning through several visual recognition experiments.

6.2 Related Work

In the late 1990s, there were several attempts to build visual recognition systems into early versions of wearable computers. Steve Mann designed and prototyped a wearable personal device that could take pictures and recognize human faces in it [14]. This wearable device was also equipped with a small head-mounted display to give textual information to users. The author stated that the system could act as a visual perception enhancer because it could provide users with real-time feedback on what they were viewing. Even though this prototype was cumbersome to wear (e.g., a set of communication units were attached to the user's body), it is considered as a pioneering example of wearable visual recognition systems. Thad Starner et al. proposed a system that recognizes the user's current behavior by analyzing video data [21]. The system utilized a hat-mounted camera to collect video streams, and then classified each single frame into pre-defined categories using a probabilistic object recognition technique. By using the results of object recognition, the authors trained a hidden Markov model (HMM) to identify three different tasks performed by the user. They also developed a visual recognition system capable of recognizing sentence-level American Sign Language selected from a 40-word lexicon [22]. They collected input video streams from both a desktop computer and from a wearable computer (namely the same device as in [21]). The experimental results showed that the system could recognize the given sign language subset with up to 98% accuracy. Finally, Antonio Torralba et al. also proposed a wearable system that accurately identified 24 different of object types in a given image [23]. First, the authors adopted an HMM approach to recognize the current location of the user. Next, they utilized the location information as a contextual cue for detecting objects from an image, based on Bayesian inference. A helmet-mounted webcam was used to collect training image data under realistic conditions in which the user walked freely around the environment.

All of the mentioned works provide useful insight and practical advice for developing visual recognition systems for wearable computers. However, this topic has not been actively studied anymore after the early 2000s. This is probably because there were no commercial camera-equipped wearables available, leading to less opportunities for research in both academia and industry. However, with the advent of Google Glass, this situation may change. For instance, researchers at Fraunhofer developed emotion recognition software for Google Glass [3]. Based on their proprietary machine learning framework SHORE, the system detects people's faces in an image taken by Google Glass, and determines their emotional states by analyzing facial expressions. Similarly, researchers in the field of affective computing connected Google Glass with custom-made smile detection software to provide users with a real-time visualization of smiling faces of people around them [4]. Thomas Way et al. designed a Google Glass application named ELEPHANT for retrieving meta information about the context (e.g., activity information) in which a picture was taken [24]. They anticipated that ELEPHANT could help people with memory impairment because it can provide contextual information when they have difficulty remembering a specific object. The authors consider using traditional machine learning algorithms such as logistic regression, support vector machines (SVMs) or Naïve Bayes to retrieve context information from an image.

Recently, researchers are trying to apply deep learning methods to wearable computers to achieve more accuracy in visual recognition systems. Recently, several companies demonstrated image classification with Google Glass [5, 18]. In order to recognize objects in an image captured by Google Glass, both utilized pre-trained deep neural networks which are deployed in their cloud. Since these works have not been published as yet, we do not know the details of their systems.

However, it seems clear that they are focused on the classification of an input image using pre-trained deep learning models, rather than on training a deep learning model for an individual user.

6.3 Personalized Visual Recognition System via Google Glass

In this section, we discuss the design and implementation of our system in detail. We first describe an overall system architecture including software/hardware specifications, and then explain the functional details of the system.

6.3.1 System Architecture

Our system is designed as a client-server model (Fig. 6.1). As a client, Google Glass collects images when instructed by the user, and sends them to the server with a specific task type (training or classification). The server then carries out the requested task and returns the results back to Glass. The server was designed to continuously train (or update) the CNN using the proposed training mechanism whenever new image data is collected by Glass. When the server completes the training task, it replaces the preexisting CNN with the newly trained one that considers the most recent images. Overall, Google Glass acts as an image collector and interface which is visible to the end user. The server performs machine learning tasks in the background, such as classifying images when needed and training new models when an object is added. We chose this architecture because Google Glass has limited computing power. To the best of our knowledge, Glass's dual-core CPU (OMAP4430) and 2 GB main memory are not sufficient to execute backpropagation for training CNNs.

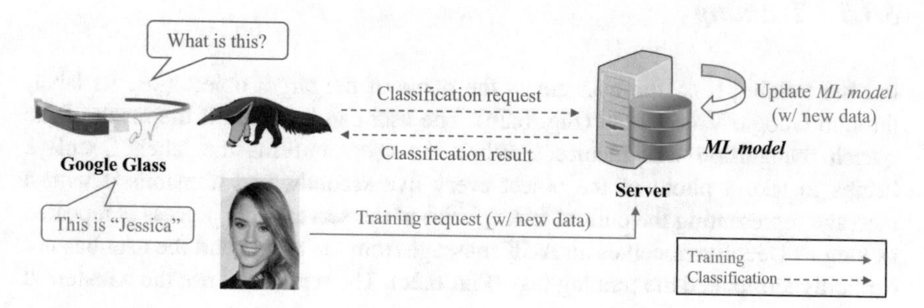

Fig. 6.1 System architecture

6.3.2 *Client*

We developed a Google Glass application (Glassware) named DeepEye, following the Ongoing Task design pattern proposed by Google. The Ongoing Task pattern is commonly used for building a basic Glassware that enables users to control their Google Glass [2]. We wrote a function for DeepEye that takes a photo periodically upon the user's command. DeepEye sends these image data and messages to the server through Java socket communication over the Wi-Fi network. We used official Google libraries such as the Android 4.4.2 (API 19) SDK and the Glass Development Kit Preview in developing DeepEye.

6.3.3 *Server*

The main purpose of the server is to quickly train deep learning models with reasonable prediction accuracy. In order to achieve this, we built a Java server on a Linux workstation equipped with a modern GPU (NVIDIA GeForce GTX 970). We then deployed an open source deep learning framework named Caffe [9] on the server. Currently, Caffe is one of the fastest CNN implementations available. If the server receives a request for a specific task from DeepEye, it executes a corresponding Caffe command (e.g., train a CNN or classify an image with a CNN) through its python interface, and returns the result.

6.3.4 *Workflow*

As discussed earlier, DeepEye has two main tasks: training and classification. Here, we describe each task step by step. When DeepEye is started, a user is asked to choose between two tasks via the Google Glass touch pad (Fig. 6.2a).

6.3.5 *Training*

For the training task, the user enters the name of the target object (i.e., its label) through Google Voice Input (Fig. 6.2b). The user can try again if the result of the speech recognition was incorrect. When the user confirms the label, DeepEye begins to take a photo of the object every five seconds, and transmits it with a message representing the current task (_train) to the server. This process is repeated as long as DeepEye receives an ACK message from the server and the user has not explicitly terminated the training task (Fig. 6.2c). The server will use the transferred

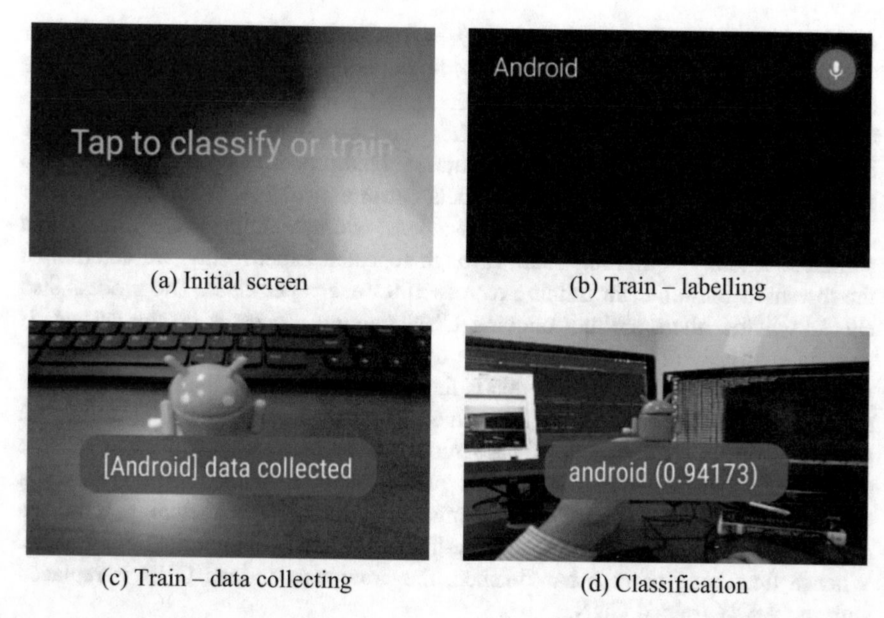

(a) Initial screen (b) Train – labelling (c) Train – data collecting (d) Classification

Fig. 6.2 Screenshots of DeepEye

image data for training a deep learning model via our proposed training mechanism which we call 'chained finetuning'.

As discussed, training deep learning models from scratch is very expensive and time-consuming. For example, training a CNN on the ImageNet dataset which contains 1.2 million images with 1,000 categories can take several weeks on a single GPU or hours/days in a distributed setting [11]. For these reasons, it is more common to retrain an already fully trained model on a new dataset to repurpose a preexisting model for different tasks [10]. For instance, after the initial retraining, we can immediately exploit the pre-trained CNN's well-learned parameters representing generic visual features like edges. Then, we can focus on updating values of parameters aimed at extracting more object-specific (high-level) features related to our own image data. This approach is known as finetuning, one kind of transfer learning algorithm. Finetuning is widely used to avoid expensive training efforts in diverse machine learning tasks [15].

Chained finetuning, the extended version of finetuning, was designed to train a new deep learning model on ad hoc additional training data. The main idea of chained finetuning is simple. To train a new model (here, CNN) for a new task, it iteratively retrains the pre-trained CNN on a newly created dataset. Suppose that there exists a CNN trained to classify an image into three user-defined categories A, B and C (CNN_ABC). If a user adds a new category D with the corresponding image data, chained finetuning then constructs a new model (CNN_ABCD) on new training data while using the old model (CNN_ABC) as a starting point. More specifically, we define a new CNN by adopting an underlying network architecture of the pre-trained CNN, but change its classification layer to have a correct number of outputs based on the given task (e.g., 4 output nodes for CNN_ABCD). Next, we can initialize parameters (weights) of the new CNN with that of the pre-trained CNN, and then progressively update the weights of the new CNN through the backpropagation algorithm on a new dataset. This process can be continued in a series whenever new types of training data became available.

Chained finetuning begins if there are at least two user-defined categories with a sufficient amount of training data. Through repeated experiments, we determined the threshold for sufficient training data as 100 images per class. The process also checks whether there are any ongoing CNN training processes on the system. If training is already in progress, it will not try to train a new model until the ongoing process has ended. Otherwise, if this is the first finetuning attempt, it trains a new model by using the pre-trained CNN named BVLC Reference CaffeNet (CaffeNet) [9]. We utilized CaffeNet as a base model because it is a publicly available pre-trained CNN that has a reasonable prediction performance on a 1,000-class object recognition task (ImageNet challenge) [11]. In any later finetuning, it trains a new model by finetuning the CNN pre-trained in the previous finetuning stage. When a finetuning process has finished, the previously trained CNN is replaced with the newly trained CNN.

6.4 Classification

Classification is relatively simple. When users choose the classification task, they take a picture of the object by clicking the Google Glass touch pad. Similar to the training task, DeepEye sends the image to the server, but with a different message (_classify). Next, the server uses the latest trained CNN to execute the Caffe classification command on the image. If no error occurs, the server sends the classification result (with probability) back to DeepEye. If DeepEye receives the result from the server, it displays them to the user through Google Glass's heads-up display (Fig. 6.2d).

6.5 Experiment 1: Person Identification

6.5.1 Overview

To validate the effectiveness of the chained finetuning mechanism, we designed and conducted a series of person identification experiments. At first, we finetuned CaffeNet so that it could identify 20 different people, rather than 1,000 different objects from a set of images. The intention was to confirm that finetuning is an effective approach for constructing a custom deep learning model for a new task.

This was important because a single finetuning step is the basic building block for chained finetuning. Second, we finetuned the previously trained CNN while adding images of a new person to the training data (i.e., chained finetuning), and evaluated the predictive power of the CNN trained in each single finetuning stage. As a result, we trained a custom CNN so that it could classify five different people. While finetuning CNNs, we tried to update the weights of the classification layer faster than that of the underlying (low-level) layers. This is because low-level layers of CNNs are supposed to extract more generic visual features (e.g., edges), and therefore they likely do not change much when presented with new data. Higher layers, in contrast, represent more class-specific characteristics (e.g., shapes) and thus need major updating with new data. Finally, we compared chained finetuning with the original finetuning approach to decide which is better for training personalized deep learning models.

6.5.2 Training Data

To gather training data, we randomly downloaded photos of 20 celebrities via Google Image Search. Using a simple shell script, we collected a maximum of 100 images for each person. We excluded some duplicate or corrupt images, and hence the number of images per class (person) was not the same (see Table 6.1). We cropped faces from original images using the OpenCV library to better gauge how well the trained CNNs identify different faces. We also augmented training data by creating additional image transformations using ImageMagick's 'convert' tool. Specifically, we created four variations of each original image through 90, 180 and 270° rotation and mirroring. We included this step to alleviate potential overfitting problems as much as possible by providing more training data without extra labelling cost (data augmentation [7, 11]). In total, our training data included 6,220 images. To measure training and test errors of the trained CNNs, we shuffled the training data and put 20% aside as a test data set.

6.5.3 Finetuning for 20-Class Person Identification

We finetuned CaffeNet for identifying 20 different face photos. By using all images described in Table 6.1 as training data, we updated all the weights in CaffeNet via backpropagation, with a maximum of 5,000 iterations. The training curves depicted in Fig. 6.3 show that the finetuned CaffeNet started to converge around the 1,000th iteration. In our training/test data set, we could not observe any serious overfitting as both training and test error show a similar pattern during the entire training

Table 6.1 Person identification—training data

No	Label	Number of images[a]	Characteristics (s)
1	Jessica Alba	68 (340)	Female, 30
2	Kate Upton	54 (270)	Female, 20
3	Scarlett Johansson	67 (335)	Female, 30
4	Emma Watson	73 (365)	Female, 20
5	Jennifer Lawrence	60 (300)	Female, 20
6	Arnold Schwarzenegger	49 (245)	Male, 60
7	Johnny Depp	63 (315)	Male, 50
8	Bill Gates	59 (295)	Male, 60
9	Kristen Stewart	80 (400)	Female, 20
10	Leonardo Dicaprio	81 (405)	Male, 40
11	Lionel Messi	55 (275)	Male, 20
12	Manny Pacquiao	51 (255)	Male, 30
13	Matt Damon	74 (370)	Male, 40
14	Michael Jackson	47 (235)	Male, 50
15	Sandra Bullock	75 (375)	Female, 50
16	Eminem	39 (195)	Male, 40
17	Steve Jobs	55 (275)	Male, 50
18	Tiger Woods	58 (290)	Male, 40
19	Tom Cruise	74 (370)	Male, 50
20	Will Smith	62 (310)	Male, 40

[a]The number in parentheses indicates the number of augmented training images

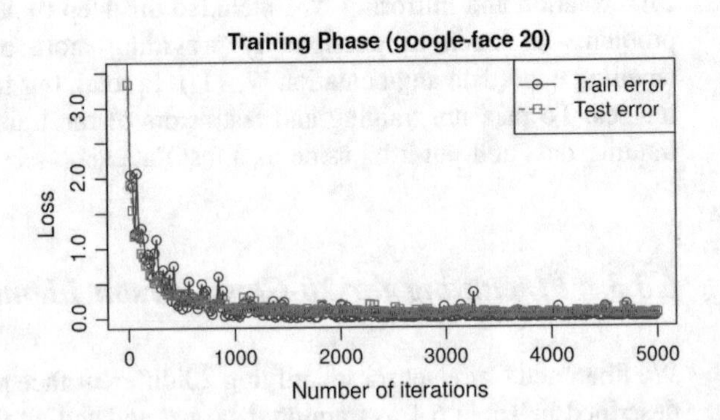

Fig. 6.3 20-Class 'Person' identification—training curves

process. For 40 consecutive tests on the 20% test data set, its average prediction accuracy was about 0.98 and its loss (error) was about 0.05. We therefore conclude that finetuning is effective for transforming a preexisting deep learning model into the new model that performs a different task.

6.5.4 *Chained Finetuning for 5-Class Person Identification*

The goal of this experiment was to train a new CNN that identifies 5 different people through chained finetuning. The experiment was conducted in the following steps. First, we finetuned CaffeNet on training images of class 1-3 so as to identify 3 different faces, and used it as a base model for chained finetuning. Next, we continued to finetune the previously trained CNN whenever a new data class was added. Two additional classes of image data were added in turn to the previous training data (class 'new-1' and 'new-2' in Table 6.2). There are 1,625 images in the training data. We shuffled and split them into 80% training and 20% test data.

Since we noticed in the previous experiment that the finetuned network converged around the 1,000th iteration, we decided to stop our individual finetuning at this point. Table 6.3 summarizes the prediction performance of the chain-finetuned CNNs (CF_CNN) on the 20% test data set. Similar to the previous experiment, the test accuracy of the finally trained CNN was nearly perfect (99%). In addition, all CNNs trained through the chained finetuning mechanism also showed promising test accuracies. Figure 6.4 displays training curves for the finetuned model, CF_CNN (5). As with the previous experiment, no serious overfitting on the training/test data sets was observed.

6.5.5 *Comparison Between Finetuning and Chained Finetuning*

One possible approach to cope with an ad hoc addition of a new data class is to train a new CNN using CaffeNet as a *fixed* base model whenever new data is added,

Table 6.2 5-Class 'Person' identification—training data

No	Label	Number of images	Characteristics (s)
1	Jessica Alba	68 (340)	Female, 30
2	Kate Upton	54 (270)	Female, 20
3	Scarlett Johansson	67 (335)	Female, 30
new-1	Alexandra Daddario	70 (350)	Female, 30
new-2	Amanda Seyfried	66 (330)	Female, 30

Table 6.3 5-Class 'Person' identification—test accuracy

Finetuned model (Number of Classes)	Base model (Number of Classes)	Test accuracy (Loss)
CF_CNN (3)	CaffeNet (1,000)	0.9885 (0.0427)
CF_CNN (4)	CF_CNN (3)	0.9956 (0.0278)
CF_CNN (5)	CF_CNN (4)	0.9969 (0.0134)

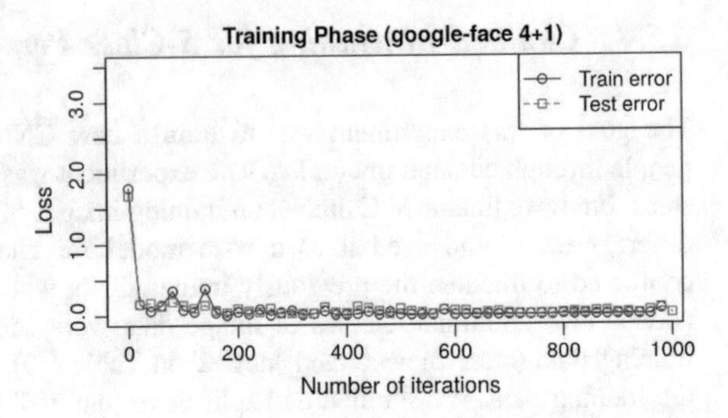

Fig. 6.4 5-Class 'Person' identification—training curves

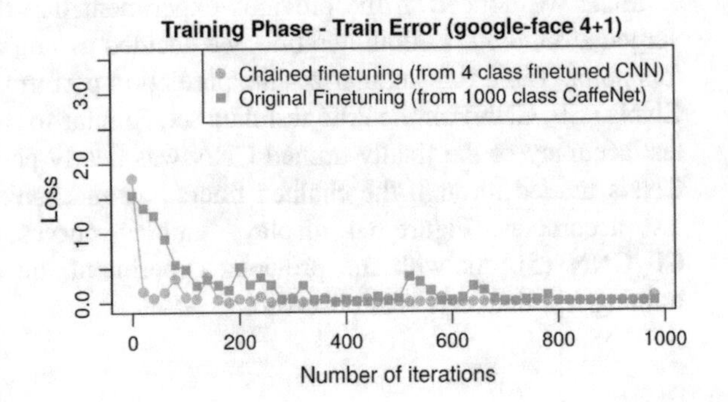

Fig. 6.5 5-Class 'Person' identification—finetuning versus chained finetuning (Training Error)

which is the original finetuning approach. To compare this approach with chained finetuning, we used original finetuning in training a CNN on training data used in the previous experiment (finetuned CNN; F_CNN). Then, we compared its prediction power with the CNN trained through chained finetuning (chain-finetuned CNN; CF_CNN). To gauge the models' prediction power more objectively, we collected an additional set of 50 images per each class. These images were downloaded from a different source (Bing image search) and never used in the training process. We used them as a validation data set for this experiment.

As shown in Fig. 6.5, the chain-finetuned CNN starts to converge about 30% earlier (after 200 iterations) than the finetuned CNN (after 700 iterations). This was expected, since chained finetuning takes advantage of what was already learned from the previous step. On both the test and validation data sets, the chain-finetuned CNN outperformed the finetuned CNN (see Fig. 6.6 and Table 6.4). However, we also noticed that the performance on validation data (validation accuracy) was lower than the test accuracy in both cases. This implies that the trained CNNs might be excessively fitted to the training data, thus having difficulties to predict outcomes for previously unseen data. We suspect that the unbalanced distribution of training data is one possible reason for this overfitting problem. There were 400 training images for class 9, but 195 images for class 16 (see Table 6.1). The model may not have been sufficiently trained for identifying class 16. For the following experiments, we tried to assign an equal amount of training data to each individual class to prevent overfitting as much as possible.

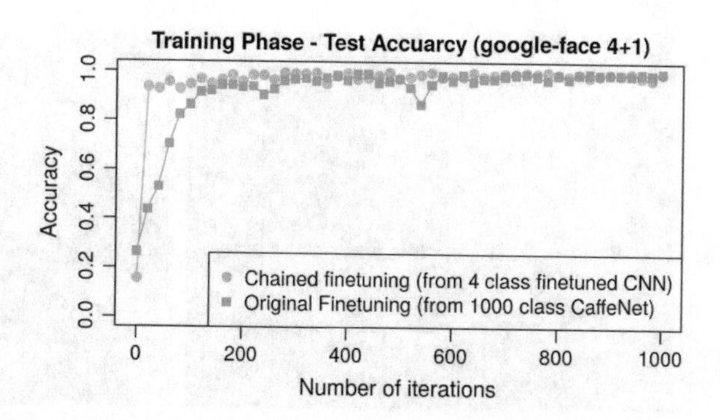

Fig. 6.6 5-Class 'Person' identification—finetuning versus chained finetuning (Test Accuracy)

Table 6.4 5-Class 'Person' identification—test and validation accuracy

Finetuned model (Number of Classes)	Test accuracy (Loss)	Validation accuracy (Loss)
F_CNN (5)	0.9656 (0.0893)	0.844 (0.7122)
CF_CNN (5)	0.9969 (0.0134)	0.88 (0.6967)

6.6 Experiment 2: Object Recognition

6.6.1 Overview

In this experiment, we aimed to evaluate the predictive power of chain-finetuned CNNs in a real-world scenario. To this end, we trained a CNN so that it can recognize 10 different types of objects from images taken by Google Glass. The ultimate aim of such a system would be to help people with memory problems to remember and recognize their personal belongings.

6.6.2 Training and Validation Data

To begin with, we chose 10 personal objects (small toy, badge, baseball cap, key, eyeglasses, pouch, food container, lotion, watch, wallet) of a member of our research team, and collected images using DeepEye and the server. To minimize the risk of overfitting, we collected the exact same amount of training data for each class, namely 100 original with 400 automatically augmented images. We also collected 30 additional images per each class as validation data. To differentiate these from the original training data, we deliberately varied the photographing conditions such as lighting, angle and background (see Fig. 6.7). Both training and validation images were taken by a single participant in a standard office setting. Even though Google Glass is equipped with a 5MP camera capable of taking 2,560 by 1,888 resolution JPG images with a file size of about 2 megabytes, we collected

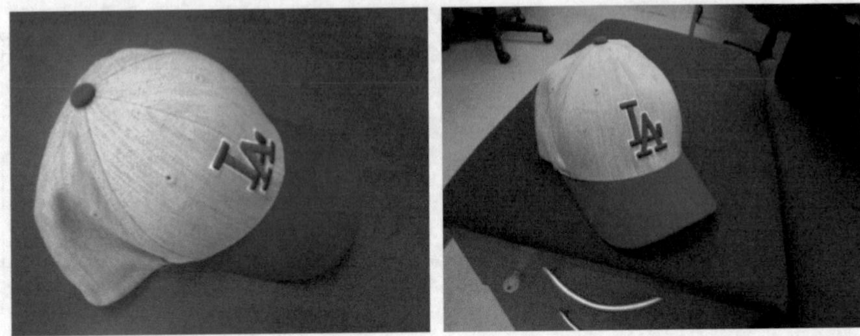

(a) Sample Image for Training (cap) (b) Sample Image for Validation (cap)

Fig. 6.7 10-Class 'Object' recognition—training and validation data

Fig. 6.8 10-Class 'Object' recognition—training curves

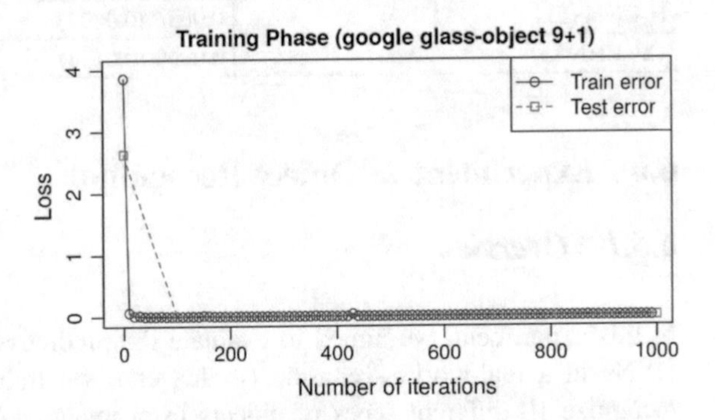

reduced-size versions of the images (1296 by 972 pixels) to avoid any network delays between DeepEye and the server.

6.6.3 *Chained Finetuning for 10-Class Object Recognition*

Regarding chained finetuning, we used the same procedures and settings as for the 5-class person identification experiment described above. The training curves in Fig. 6.8 show that the finetuned CNN starts to rapidly converge at around 100 iterations. Compared to all previous person identification experiments, this one had near-perfect test accuracy, probably because the classification task was easier. The objects used in this experiment had vastly different shapes and appearances (e.g., cap vs. wallet), so that the model could identify them with high confidence. In contrast, the differences between faces of the same gender and age are subtle (e.g., class 4: Emma Watson and class 9: Kristen Stewart in Table 6.1). This could have led to some confusion telling them apart.

Table 6.5 10-Class 'Object' recognition—validation accuracy

Finetuned model (Number of Classes)	Base model (Number of Classes)	Validation accuracy (Loss)
CF_CNN (3)	CaffeNet (1,000)	0.99 (0.0212)
CF_CNN (4)	CF_CNN (3)	0.99 (0.1819)
CF_CNN (5)	CF_CNN (4)	0.99 (0.0555)
CF_CNN (6)	CF_CNN (5)	0.99 (0.0462)
CF_CNN (7)	CF_CNN (6)	0.99 (0.0454)
CF_CNN (8)	CF_CNN (7)	0.96 (0.2696)
CF_CNN (9)	CF_CNN (8)	0.94 (0.2319)
CF_CNN (10)	CF_CNN (9)	0.97 (0.116)

Table 6.5 summarizes the measured prediction power of all chain-finetuned CNNs on the validation data set. For up to 7 different objects, the trained CNNs showed a near-perfect performance in recognizing objects without serious overfitting concerns. However, the validation accuracy was slightly diminished, as the number of object classes increased from 8 to 9. It may be improved if we collect additional training images of the objects, and train a new CNN at the next fine-tuning stage. The validation accuracy of the final trained CNN was 97%, with a loss of 0.116. It took approximately 7 min to train this model on our GPU platform.

6.7 Discussion and Future Work

We demonstrated the feasibility of our proposed visual recognition system that uses Google Glass. Yet, it still has some issues that need to be overcome before widespread use.

Google Glass emits a lot of heat when it continuously utilizes the camera function. According to [13], a single camera usage heats Google Glass 28 °C above the surrounding temperature. In the worst case (video chatting), Google Glass's surface temperature increased up to 50 °C within 13.3 min. Because Google Glass is in direct contact with the skin, the heated surface may lead to discomfort and potentially even health risks for users. Therefore, users may have trouble collecting at once a large volume of images (more than 100) via Google Glass. The authors also measured the energy consumption of Google Glass for various tasks. To take a single photo, Google Glass consumes 2,927 mW for 3.3 s. Considering its battery capacity, users can take fewer than 800 images on a single charge. Like the heat problem, this may prevent users from taking sufficiently many images to build their own deep learning models. We expect Google to fix these issues in the next generation of Google Glass.

At this moment, there exists no large-scale image dataset collected from wearable computers such as Google Glass. Therefore, we generated the custom dataset using DeepEye in our experiments, and utilized it for testing the proposed training mechanism. To thoroughly verify its effectiveness, we should investigate whether our mechanism also works well for more complex image classification problems (e.g., 100-class object recognition). Thus, we are considering distributing DeepEye to a group of Google Glass users, and to collect abundant image data from their everyday lives. Furthermore, we need to tackle any potential overfitting problems in training personalized deep learning models. For this, we applied a neural network regularization technique named dropout [20] to DeepEye's training mechanism. Dropout forces neural networks to learn several independent representations of identical input-output pairs, by randomly disabling some neurons (nodes) in a given layer. For all experiments described above, we used a fixed dropout rate of 0.7 for fully-connected layers. Therefore, it is worth investigating the optimum dropout rate for training more complex models through chained finetuning.

While often overlooked, privacy is an important concern [6, 8]. For the person identification task, users need to take photos of people around them (mostly, friends and acquaintances). We assumed that they would ask them for their permission before taking a photo. However, there are no user interfaces or mechanisms in our system advising them to do that. The system may invade privacy if it collects photos of people without their consent. To find the best way to prevent possible privacy invasions, we need to collect users' opinions on, and/or reactions against the system. Also, it is necessary to prevent unauthorized access to user-generated image data and the trained models, as they may reflect a user's very personal behavior and interests.

Finally, we believe that even users who are not tech-savvy should have little difficulty using our system because they are only asked to perform a few simple operations via Google Glass (e.g., image labelling through Google Voice Input). However, we need to verify the usability of the system with target users who have special needs. To explain, we need to qualitatively and quantitatively assess the usability of the system to people with memory or visual impairment, possibly including their caregivers. Their feedback may allow us to improve our user interface, so that our system will work in a more user-friendly way. Additionally, a longitudinal study might need to be conducted to verify whether our system can have a positive influence on their lives and medical conditions.

6.8 Conclusion

In this paper, we designed and implemented a novel wearable system which builds personalized deep learning models for recognizing objects of interest to a user. To the best of our knowledge, this is the first attempt to train deep learning models for *personalized* visual recognition, via camera-equipped wearable computers like Google Glass. The proposed system works as a client-server model: Google Glass (client) collects images from a user's everyday life and sends them to a GPU-equipped Linux server. The server then trains a deep convolutional neural network (CNN) on the user-specific image data. To efficiently update the pre-trained network on newly-added images, we proposed a simple training mechanism called chained finetuning. As a variant of conventional finetuning, it is effective in terms of prediction power and training efforts in continuously training (or updating) a personalized deep learning model. In a custom 10-class object recognition task, our system took 7 min to train a personalized CNN on our GPU platform, and showed a 97% classification accuracy without serious overfitting. Considering the training time and the model's prediction power, we believe our system can become a feasible intelligent personal assistant. Future work will mainly focus on the testing of the proposed system with more users and harder tasks. It will also include privacy impact assessments and a verification of its effectiveness in improving users' cognitive abilities.

Acknowledgements Part of this work was done while Hosub Lee was a summer intern at Samsung Research America, Mountain View, CA. The authors would like to thank reviewers for their valuable comments on earlier versions of this paper.

References

1. Ciresan, D., Meier, U., Schmidhuber, J.: Multi-column deep neural networks for image classification. In: 2012 IEEE Conference on Computer Vision and Pattern Recognition (CVPR), pp. 3642–3649. IEEE (2012)
2. Google Developers: Ongoing task pattern. (2015). https://developers.google.com/glass/develop/patterns/ongoing-task
3. Fraunhofer: Fraunhofer IIS presents world's first emotion detection app on Google Glass (2014). http://www.iis.fraunhofer.de/en/pr/2014/20140827_BS_Shore_Google_Glas.html
4. Hernandez, J., Picard, R.W.: SenseGlass: using Google Glass to sense daily emotions. In: Proceedings of the Adjunct Publication of the 27th Annual ACM Symposium on User Interface Software and Technology, pp. 77–78. ACM (2014)
5. Hof, R.: First image recognition app coming soon to glass (2014). http://www.forbes.com/sites/roberthof/2014/02/26/first-image-recognition-app-coming-soon-to-google-glass/
6. Hong, J.: Considering privacy issues in the context of Google Glass. Commun. ACM **56**(11), 10–11 (2013)
7. Howard, A.G.: Some improvements on deep convolutional neural network based image classification (2013). arXiv:1312.5402
8. Hoyle, R., Templeman, R., Armes, S., Anthony, D., Crandall, D., Kapadia, A.: Privacy behaviors of lifeloggers using wearable cameras. In: Proceedings of the 2014 ACM International Joint Conference on Pervasive and Ubiquitous Computing, pp. 571–582. ACM (2014)
9. Jia, Y., Shelhamer, E., Donahue, J., Karayev, S., Long, J., Girshick, R., Guadarrama, S., Darrell, T.: Caffe: Convolutional architecture for fast feature embedding. In: Proceedings of the 22nd ACM International Conference on Multimedia, pp. 675–678. ACM (2014)
10. Karayev, S., Trentacoste, M., Han, H., Agarwala, A., Darrell, T., Hertzmann, A., Winnemoeller, H.: Recognizing image style (2013). arXiv:1311.3715
11. Krizhevsky, A., Sutskever, I., Hinton, G.E.: ImageNet classification with deep convolutional neural networks. In: Advances in Neural Information Processing Systems, pp 1097–1105 (2012)
12. LeCun, Y., Bengio, Y., Hinton, G.: Deep learning. Nature **521**, 436–444 (2015). doi:10.1038/nature14539

13. LiKamWa, R., Wang, Z., Carroll, A., Lin, F.X., Zhong, L.: Draining our glass: an energy and heat characterization of Google Glass. In: Proceedings of 5th Asia-Pacific Workshop on Systems, p. 10. ACM (2014)
14. Mann, S.: Wearable computing: a first step toward personal imaging. Computer **30**, 25–32 (1997). doi:10.1109/2.566147
15. Pan, S.J., Yang, Q.: A survey on transfer learning. IEEE Trans. Knowl. Data Eng. **22**, 1345–1359 (2010). doi:10.1109/TKDE.2009.191
16. Schmidhuber, J.: Deep learning in neural networks: an overview. Neural Netw. **61**, 85–117 (2014). doi:10.1016/j.neunet.2014.09.003
17. Sermanet, P., Eigen, D., Zhang, X., Mathieu, M., Fergus, R., LeCun, Y.: OverFeat: Integrated recognition, localization and detection using convolutional networks (2013). arXiv:1312.6229
18. Simonite, T.: A Google Glass app knows what you're looking at (2013). http://www.technologyreview.com/view/519726/a-google-glass-app-knows-what-youre-looking-at/
19. Simonyan, K., Zisserman, A.: Very deep convolutional networks for large-scale image recognition (2014). arXiv:1409.1556
20. Srivastava, N., Hinton, G., Krizhevsky, A., Sutskever, I., Salakhutdinov, R.: Dropout: a simple way to prevent neural networks from overfitting. J. Mach. Learn. Res. **15**, 1929–1958 (2014)
21. Starner, T., Schiele, B., Pentland, A.: Visual contextual awareness in wearable computing. In: Second International Symposium on Wearable Computers, pp. 50–57. IEEE (1998)
22. Starner, T., Weaver, J., Pentland, A.: Real-time American sign language recognition using desk and wearable computer based video. IEEE Trans. Pattern Anal. Mach. Intell. **20**, 1371–1375 (1998). doi:10.1109/34.735811
23. Torralba, A., Murphy, K.P., Freeman, W.T., Rubin, M.A.: Context-based vision system for place and object recognition. In: Ninth IEEE International Conference on Computer Vision, pp. 273–280. IEEE (2003)
24. Way, T., Bemiller, A., Mysari, R., Reimers, C.: Using Google Glass and machine learning to assist people with memory deficiencies. In: Proceedings on the International Conference on Artificial Intelligence (ICAI), pp. 571–577 (2015)
25. Zeiler, M.D., Fergus, R.: Visualizing and understanding convolutional networks. In: European Conference on Computer Vision, pp. 818–833. Springer International Publishing (2013)

Chapter 7
Intelligent Personal Assistant for Educational Material Recommendation Based on CBR

Néstor Darío Duque Méndez, Paula Andrea Rodríguez Marín and Demetrio Arturo Ovalle Carranza

Abstract Personal assistants are focused on helping users with various tasks in the daily management, as they anticipate their needs and learn with their interaction. An intelligent personal assistant is a software agent that can perform actions requested by a user and can access to information from remote sources, based on requirements or user profile. Moreover, intelligence personal assistants can be considered as a special case of recommendation systems since they are used in web searches. Thus, the personal assistant interacts and represents users to choose relevant items according to their needs and preferences. This work proposes an intelligent personal assistant aimed to support users for selecting educational material from learning objects repositories. In this regard, a recommendation system was implemented based on the artificial intelligence technique known as CBR. The possibility of taking advantage of previous results of students with similar characteristics allows to improve the relevance of the materials for each particular student. The results of the functional tests are satisfactory.

Keywords Personal assistant · Recommender system · Case-based reasoning · Learning objects

N.D. Duque Méndez
Universidad Nacional de Colombia Sede Manizales, Manizales, Colombia
e-mail: ndduqueme@unal.edu.co

P.A. Rodríguez Marín (✉) · D.A. Ovalle Carranza
Universidad Nacional de Colombia Sede Medellín, Medellín, Colombia
e-mail: parodriguezma@unal.edu.co

D.A. Ovalle Carranza
e-mail: dovalle@unal.edu.co

A. Costa et al. (eds.), *Personal Assistants: Emerging Computational Technologies*, Intelligent Systems Reference Library 132, DOI 10.1007/978-3-319-62530-0_7

7.1 Introduction

The objective and the role of a personal assistant can be varied, such as to help to the user with daily management, scheduling of meetings, correspondence, answering messages, etc. They born with the need of the users to find the information among the growing and complex sources of information. They can be seen as an information agent that differs from others because of their interaction with users and their information of a dynamic nature [1].

An intelligent personal assistant can be seen as a recommendation system that interacts with the user to learn and continuously modify the user's profile, the wizard can also represent the preferences and choose which items should be relevant to the user. In the area of education, the personal assistant supports the student through the suggestion of educational resources that fit or adapt their characteristics, this is the general objective of the recommendation systems.

Personal assistants can be considered as adaptive intelligent agents that are able to adapt themselves and can be applied to learning personal assistants on the Web [2]. Adaptive information agent systems employ learning techniques to adapt to one or all of the following: users, agents, and the environment. Examples include personal assistants for information searches on the Web or collaborating information agents that adapt themselves as a system in changing environments [3].

Likewise, a Recommendation Systems (RS) is defined as a piece of software that facilitates users to discern more relevant and interesting learning information [4]. RS are a tool aims at providing users with useful information results searched and recovered according to their needs, making predictions about matching them to their preferences and delivering those items that could be closer than expected [5]. In the educational application, the descriptive metadata of the educational resources is used, to make recommendations that are adapted to the preferences and needs of the students [6]. The students need to know that there are others learning models and compare if the personal model that they use is the best. In addition, a learning style is a description of a process, or of preferences. Any inventory that encourages a learner to think about the way that he or she learns is a useful step towards understanding and hence improving, learning [7]. Besides, is necessary that students "learn to learn" and teachers should recognize the individual differences of their students to customize their education. A student's learning style is how they process information [8].

Customization in the virtual environment from student recognition is part of the original expectations in the use of Information and Communication Technologies (ICT) in educational systems.

Artificial intelligence techniques offer great potential and good results. The state of the art shows that there is a wider application of these techniques in various stages of the educational process and with favorable results [3]. The approach proposed in this paper is based on Case-Based Reasoning (CBR), which stems from the idea of applying similar solutions to similar problems and applies these principles for obtaining successful resources for students with equivalent features. CBR technique has been used successfully in different works to support the educational processes [9].

For the use of the technique is need defining the variables that compound one case, your weighing, and the objective feature; the measuring method of the similarity and the strategies for each phase of application of the technique. Aamodt says that, as for AI in general, there are no universal CBR methods suitable for every domain of application. The challenge in CBR is elsewhere to come up with methods that are suited for problem-solving and learning in particular subject domains and for particular application environments [10].

The research that supports this work tries to solve the initial question of how to design and construct a personal intelligent assistant for LO recommendation, of which appears an emergent question about the technique of artificial intelligence that fulfilled the end and was simple enough for its easy implementation and execution with modeling requirements. The evaluation of alternatives and the experience in [11] previous allowed to conclude that CBR is adequate to the proposed objective.

The methodology for designing an intelligent personal assistant, which allows making recommendations of educational resources using CBR, requires first determining the elements in the student profile and the metadata of the learning object (construction of the cases). Second, define the metrics for the recovery of the most similar cases (compare with the new problem), and finally, determine the update of the case base (add new case).

In this paper, we propose a model that was validated with a case study that shows that it is possible to deliver adapted learning objects using a personal assistant for CBR based recommendation. It also allowed the conclusion that the implementation of CBR is promising and shows its possibilities in such type of problems, and the mechanism used to solve the problem of non-existing cases allowed obtaining new cases dynamically, while always delivering LO to students according to their profile and the metadata of these resources, constituting a support in teaching and learning processes.

The rest of the paper is organized as follows: Sect. 7.2 is a conceptual revision, Sect. 7.3 reviews related work in the area about the personal assistant, CBR, and learning object recommendation. An overview of our intelligent personal assistant for the educational material recommendation based on CBR is presented in Sect. 7.4. Section 7.5 explains the model implementation. Finally, Sect. 7.6 summarizes the contributions of the paper and proposes future work.

7.2 Preliminaries

In this section, we present the main concepts related to an intelligent personal assistant for the educational materials recommendation, using the artificial intelligence technique: case-based reasoning.

7.2.1 Learning Objects (LO), Learning Objects Repositories and Repository Federation

Learning Objects (LO): They are digital entities that can be used for learning, education or training, usually delivered via the Internet and designed to be used and reused in multiple educational settings [12]. In general terms, an LO can have any type of content, with different formats for students to learn about the issues in question.

The Learning Objects are a digital material with different granularity, which can be used for educational purposes based on an intentionality defined implicitly or explicitly by educational objectives and they contain metadata that allows its description and retrieval, which facilitates its reuse and adaptation to different environments [13]. For example, a learning object can be a slide show of a specific topic, so it can be an image or a video. Anything that helps the student learn a subject in a virtual environment.

The increase of available digital information, ubiquitous networks, and advances in technology have allowed access to thousands of educational resources and the development of resources to support the teaching-learning processes, it is there that the objects of learning as entities Digital resources that have metadata that describes these resources and allows them to search and access. Several metadata management standards have been proposed, including IEEE-LOM, Dublin-Core, and OBAA, which focus on specifying resource information [13].

Learning Object Repositories allow the organization of these digital resources oriented to education, facilitating their search and retrieval. In addition, they allow the storage, search, and recovery of multiple LO, increasing the possibility of reuse in multiple educational contexts. There are different types of repositories according to the way in which they store the resources and their respective metadata, but in general, all are oriented to facilitate access to educational materials making a great contribution in the teaching-learning processes [13].

The number of Learning Object Repositories that are being built to support teaching-learning processes is increasing, but it is necessary to provide users with effective ways to search for LO, saving time and resources required to search each repository independently. In response to this arise the Learning Object Repository Federations, providing a single point of access to users and a greater possibility of tools and services [13].

7.2.2 Case-Based Reasoning (CBR)

Case-Based Reasoning (CBR) is a technique of Artificial Intelligence that tries to reach the solution to new problems in a similar way as humans do, using the experience gained so far in similar events to make decisions in future similar cases [14].

CBR solves new problems by adapting similar solutions that were used to solve old problems to new problems. Cases are used in CBR as a mechanism of representation. A case describes one particular diagnostic situation and records several features and their specific values occurred in that situation [15].

A case is composed of three elements: The description of the problem, which may be formed by several characteristics with different weights; the solution applied, which corresponds to the response given by the system; and the solution result, which indicates whether this was appropriate or not after application [16].

Conceptually CBR is commonly described by the CBR-cycle made up of four actions: retrieve, reuse, revise and retain. Retrieval is the action of finding and returning cases similar to the one under analysis. Reusing is the action of adapting the solution retrieved so that it adapts to the new problem. Revision is the action of assessing the solution in terms of the current case, evaluating its effectiveness, and possibly reformulating it based on knowledge of the domain. This happens when the user does not agree with the solution and asks for a new case, defining which input features are relevant and what the correct fault and time to fault should be for this new case. Retention is the action of storing a newly recognized the case in memory, for future use [15]. Figure 7.1 shows the phases for the application of the CBR technique.

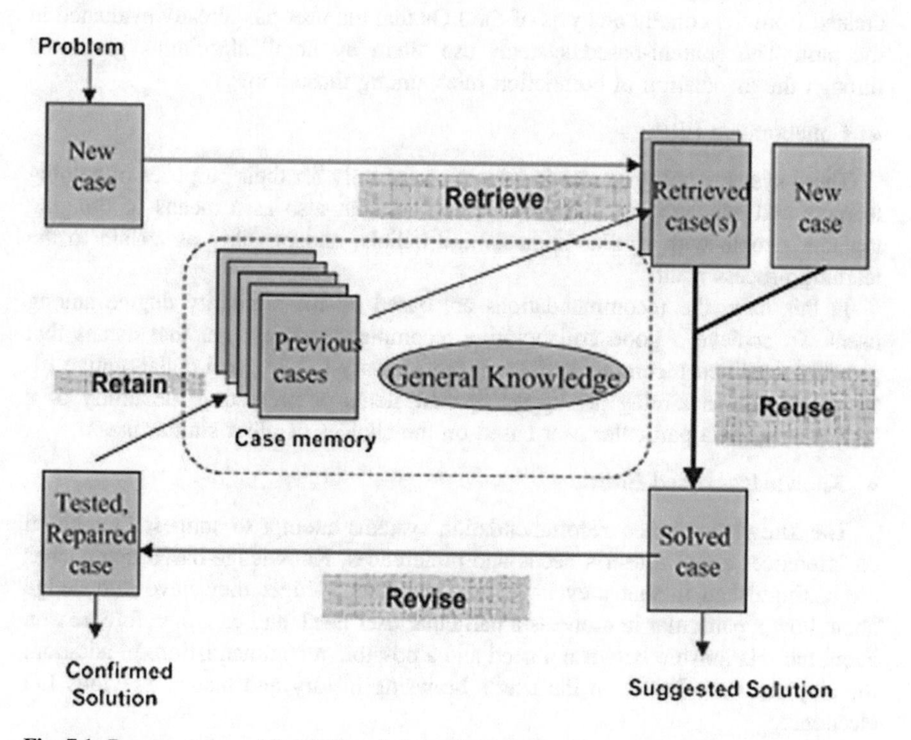

Fig. 7.1 Representation of the CBR technique phases [18]

Case-Based Reasoning Systems (CBR) requires a solid building process and a good evaluation of the case memory. Building processes mainly focus on defining (1) how the cases (instances) are stored in the case memory, (2) a retrieval strategy to obtain the most similar cases from the case memory and (3) how the solution is built from this knowledge [17].

In summary, CBR systems provide new solutions by the analogy of past design situations, based on an adaptation of the previously selected solutions [18].

7.2.3 Recommender Systems

Recommender Systems (RS) aims to provide users with search results close to their needs, making predictions of their preferences and delivering those items that could be closer than expected [5, 19]. In the context of LO, Educational Recommender Systems (ERS) deliver educational materials closer to learning needs, characteristics, and preferences for students. ERS aim to support learning/teaching process.

There are several techniques of RS as follows [20, 21]:

- Content—based ERS:

In this kind of system, the recommendations are made based on the user's profile created from the content analysis of the LOs that the user has already evaluated in the past. The content-based systems use "item by item" algorithms generated through the association of correlation rules among those items.

- Collaborative ERS:

These systems hold promise in education not only for their purposes of helping learners and educators to find useful resources, but also as a means of bringing together people with similar interests and beliefs, and possibly as an aid to the learning process itself.

In this case, the recommendations are based on the similarity degree among users. To achieve a good collaborative recommendation system that means that provides qualified recommendations, it is necessary to use good collaborative filtering algorithms aiming at suggesting new items or predicting the utility of a certain item for a particular user based on the choices of other similar users.

- Knowledge- based ERS:

The knowledge-based recommendation systems attempt to suggest LOs based on inferences about a user's needs and preferences. Knowledge-based approaches are distinguished in that they have functional knowledge: they have knowledge about how a particular item meets a particular user need, and can, therefore, reason about the relationship between a need and a possible recommendation. In addition, these systems are based on the user's browsing history and his/her previous LO elections.

- Hybrid Recommender Systems:

The hybrid approach seeks to combine the techniques ERS in order to complete their best features and thus make better recommendations. The proposed hybrid filtering approach transparently creates and maintains user's preferences.

To construct a hybrid recommendation system, you must have at least two techniques to be combined, Burke describes the following different methods [22]:

Weighted: The score of different recommendation components are combined numerically.
Switching: The system chooses among recommendation components and applies the selected one.
Mixed: Recommendations from different recommenders are presented together.
Cascade: Recommenders are given strict priority, with the lower priority ones breaking ties in the scoring of the higher ones.
Feature combination: Features derived, from different knowledge sources, are combined together and given to a single recommendation algorithm.
Feature augmentation: One recommendation technique is used to compute a feature or set of features, which is then part of the input to the next technique.
Meta-level: One recommendation technique is applied and produces some sort of model, which is then the input used by the next technique

7.2.4 Student Profile

For a RS deliver tailored results they need profiles that store the information and the preferences of each user [23]. The student profile stores information about the learner, its characteristics, and preferences. To handle a user profile can be used to support a student or a teacher in the LO selection according to its personal characteristics and preferences [24].

Some research presents the learning style as the most important feature for the delivery of educational resources [8, 25]. The learning style can be defined as the preferred strategies for capturing new information and how to use it in its environment. There are different models to represent a student's learning style.

7.3 Related Works

There are currently, related to the intelligent personal assistant, other research work on educational material recommendation, and the other hand some works related to the use of CBR to support educational customized processes are presented.

In [1] proposed to use intelligent agents that help users find information tailored to their tastes and preferences. These agents perform some filtering, selection and ordering tasks to present the information to the user who requires it.

In [2], they investigated a model and design for an intelligent agent system, which helps the user in a user-friendly fashion. This intelligent agent system was modeled in two different applications: the intelligent agent system for pattern classification and the intelligent agent system for bank asset management modeling.

Salehi et al., use genetic algorithms and realize two processes of recommendation, the first of them is the explicit characteristics represented in a matrix of preferences of the student. The second recommendation is implicit weights to educational resources that are considered as chromosomes in the genetic algorithm to optimize them based on historical values [26].

The authors deliver educational materials adapted to the user profile, combining several types of filtering with the available information about objects and users. The first method pre-selects the learning objects repository, using a search based on metadata, then those objects passed by other filtering processes to obtain a final list which will be which best suits the user in this work combines several filter criteria: based on content, collaborative activity, and demographics [27].

Klusch [28], presents information agents as a special class of software agents, among which are personal assistants who help users in the acquisition and management of information, serve to assist the user by adapting dynamically to changes in user preferences [28].

In [29] the study of personalized recommendation systems with CBR involves four interrelated aspects: the representation and the organization of the personalized recommendation on case-based reasoning, construction and maintenance of multiple cases library of personalized recommendation, judging of the similarity of personalized recommendation case and methods of retrieval, and the combination of personalized recommendation of Case-based reasoning.

In [9] is shown PeCoS-CBR, the tool based on CBR for modeling and implementation of systems for generating customized courses. Determining the structure of the Cases, the way in which recovery, adaptation, evaluation and disposal in the Case Base are made, was successful and allowed delivering educational material to students individually.

Da Rocha, Pereira dos Santos Jr., and Michelle presented a learning environment to support teaching using neural networks and CBR. The neural networks distribute the amounts of multimedia objects in the presentation of material and the role of CBR is adapting the interface according to the user's level of knowledge and is employed to select the most appropriate materials for the students, based on their profile (new case) and the characteristics of other learners (initial case-base). The similarity measure used was the nearest neighbor. They conclude that the adaptation occurs with the use of CBR, allowing learning improvement [30].

In [31], a model which integrates the case-based reasoning paradigm and the Intelligent Teaching-Learning Systems is proposed, the model favors the design of these systems by users not necessarily experts in the informatics field, taking into account the ease of use of the technique and naturalness of the case-based approach.

They show the feasibility of using a technique for the student model the reasoning based on the cases, in place of the other existing alternatives. Based on that case-based systems seem to be useful in all kinds of situations, have great versatility in student modeling and are a powerful tool for making inferences.

The author in [31] state that the adoption of educational agents and Artificial Intelligence can offer new answers to the needs of each student and provide a more effective collaboration on virtual learning environments. The learning experience of each student can be adapted to other students with the same characteristics. The article poses the approach of case-based reasoning for learning systems based on Adaptive Web used to adapt the contents of e-learning and their contexts according to the student's learning style and individual needs.

In [32] is affirmed that CBR, as a learning model, search to accumulate experiences with a succession of real cases and to properly index these experiences for later retrieval and the power of the reasoning activities through the access to old cases provides a potential instructional practice in problem-solving. The proposed CBR process emphasizes the performance of analogical reasoning and the feedback of evaluation in order for a case-based reasoner to learn its lessons while adding a new experiential episode of success or failure to its memory.

Smyth says that case-based recommender's implement a particular style of content-based recommendation that is very well suited to many item recommendation scenarios. They rely on items being represented in a structured way using a well-defined set of features and feature values. Case-based recommenders borrow heavily from the core concepts of retrieval and similarity in case-based reasoning. Items are represented as cases and recommendations are generated by retrieving those cases that are most similar to a user's query or profile [33].

Previous paragraphs allows concluding that applying CBR to problems of recommendation in educational systems involves conceptualizing, designing and implementing a multi-stage model: definition and store of the relevant characteristics of learners for customization purposes; feature or item type an offer in terms of CBR; definition of techniques or algorithms for recovery of the Cases that are closest to the present ones; selection of cases to be applied; review and adaptation of the proposed solution and storage of the new solution as part of a new case.

This work is differentiated by the incorporation into a personal intelligent assistant, as a recommendation system. This proposal aims to show the possibilities that are opened with the use of personal assistants to recommend educational resources.

7.4 Proposed Model

The intelligent personal assistant for the educational material recommendation based on CBR has the aim deliver learning objects adapted to the user profile. The proposed model consists of three principal components: the first one is the user profile, they store main characteristics and preferences of the student; the second one is the Intelligence personal assistant for recommended learning objects, this assistant is the core the proposal and is based in CBR and apply the phases for delivered adapted educational materials; and the last one component is the repository federation, this store the learning objects (resource and their metadata), for making the recommendation process.

The proposed model is generic and therefore can be applied to a different context, Fig. 7.2 presents the proposed model. The Fig. 7.3, present the process execute for the personal assistance.

7.4.1 CBR Stages in Intelligent Personal Assistant to Recommend Educational Resources

Initially, the case structure and weighting of the characteristics are defined. The features are educational level, learning style, language preference, format preference and interactivity level preference. In order to give different relevance to the features, weights are assigned and taken to account when calculating the overall similarity between the cases. The weights are assigned according to the experience in previous work of the research group and in addition, they are assigned by the relevance that each feature has for the work, as follows: level education (0.3), learning style (0.3), language preference (0.2), format preference (0.05) and interactivity level preference (0.15).

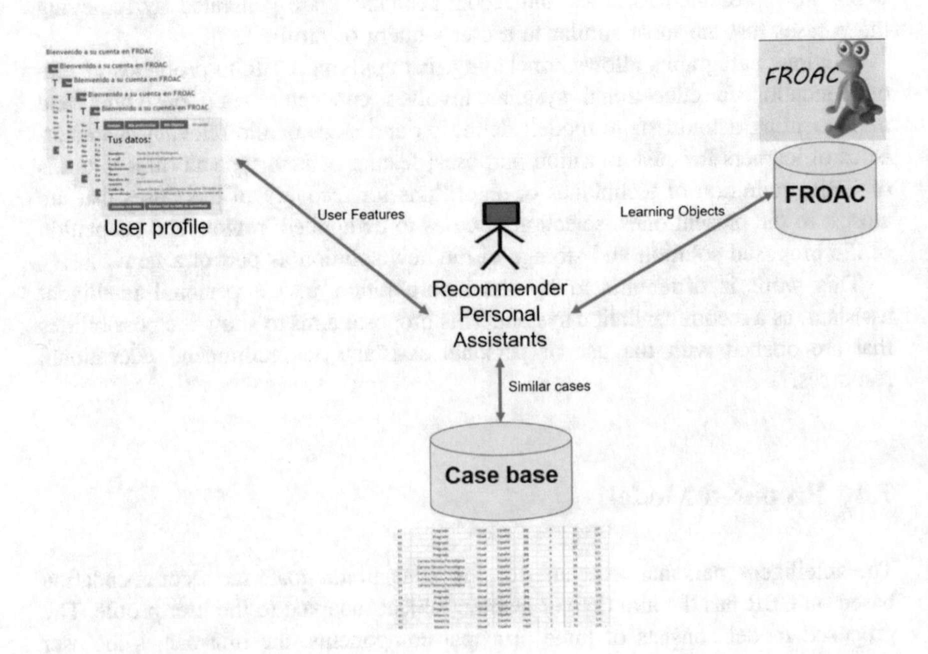

Fig. 7.2 Proposed model

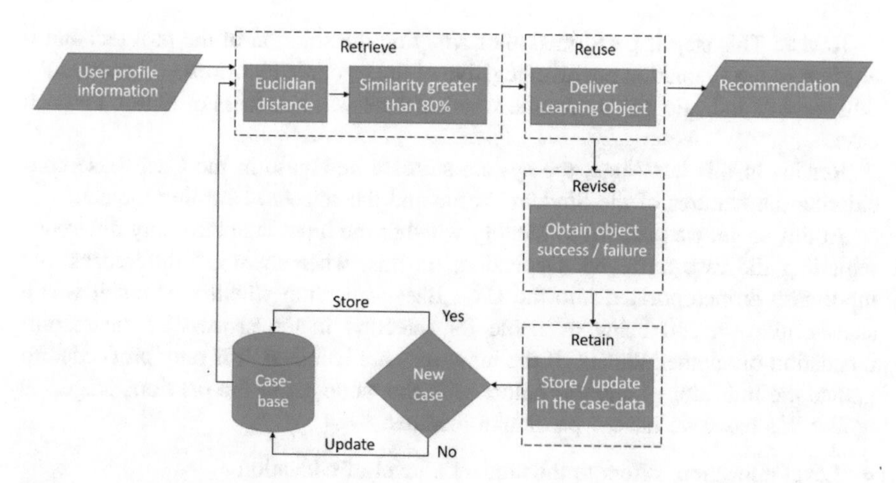

Fig. 7.3 Recommender personal assistants process

Later the ordered phases are executed.

Retrieval: The first thing that is done at this stage is to identify the characteristics of the student (Case Problem). After defining the problem, we proceed with the search task in the Case Base, calculating the level of similarity, where are identified those cases that have a level of similarity greater than 80%. The retrieval algorithm is selected from the K-nearest neighbors, where the case is retrieved considering the largest sum of the weights of each of the characteristics matching the new case.

For the calculation of the similarity of the predictive features associated with the interactivity level preference, the Euclidean distance is used, where a smaller value represents a greater similarity between the analyzed values. For the other characteristics, the Boolean metric is applied, identifying whether or not the case matches with the case analyzed. For the calculation the global similarity previously the results are normalized.

After obtaining the global similarity between the case in analysis and the record in Case Base, the most similar is identified, that is, the one with a higher overall similarity value. If the level of similarity is equal to several these are selected. From the selection of one of the cases, the corresponding Learning Object is identified, which will be the educational material to be delivered to the student.

Reuse: After the identification of the most similar cases and the selection of one of these cases, the solution, in this case, can be used directly as a solution to the new problem. Therefore, the LO that had been given to a student with a similar profile to that of the current student is presented to the latter.

The adaptation task is run on the CBR systems when modifications to the solution are necessary so that it can be applied to the new case. To make the recommendation, the solution is specifically an LO to cover the user preferences, so that an initial change is not necessary, implying that the adaptation process is done in the next stage when the case is discussed and it is determined whether the proposed solution was adequate or not.

Revise: This step is performed after applying the solution of the problem and it consists in the evaluation positive/negative of the resource by the user. These results will serve as indicators to determine whether there was a success or failure for each case.

Retain: In this last stage, the system stores a new case in the Case Base considering the features of the Student Profile and the selected Learning Object.

At this stage, we proceed to identify whether the input data have any difference regarding the case retrieved. Depending on this, when there are differences, the input case is incorporated into the Case Base indicating whether or not it was a successful case, and being available for selection in the process of the recommendation of another student. If the input data are identical, it is only proceeded to update the indicator of success/failure with the value set in the previous stage.

For this case, we have 5 predictive features:

- Level education: refers to the student's level of education
- Learning style: the way the student learns
- Language preference: language of preference of learning objects
- Format preference: format of preference of learning objects
- Interactivity level preference: Preference for interaction with the system

The output features or objectives features are:

- id_LO: identification of the recommended object
- Successes/Failures: this characteristic has the information of the successes and failures that the LO has had. Students can evaluate objects on a scale of 1 to 5, where 1 is not liked/served and 5 is liked/served. The successes are taken as the qualification 4 and 5, and the failures are the evaluations 1 and 2.

7.5 Experiments and Results

Taking the proposed model for learning objects recommendation supported by CBR and taking into account the different elements described, a prototype was built, which allowed verifying the validity of the proposal. The development was done with the PHP programming language and the PostgreSQL database manager.

In the system, the student can register; the registration page asks the student for explicit information such as personal data (name), academic data (learning style, level education) and preferences (language, format, interactivity)

The student-users have the options of My Profile, where they can view and modify their personal data, preferences and learning style.

The system delivers to the student a list of recommended LOs from similar cases. The recommendation process starts from the search criteria that can be expressed by keywords.

Below, Table 7.1 shows some the Case Base after a series of interactions of the students, with a total of 153.

Table 7.1 Fragment of case base

Id_case	Level education	Learning style	Language preference	Format preference	Interactivity level preference	Id_LO	Successes/failures
1	Básica Primaria	Lector	Español	jpg	4	52	1/0
2	Educación Superior	Lector	Español	mp4	2	40	1/0
3	Educación Superior	Lector	Español	mp4	2	41	1/0
4	Educación Superior	Lector	Español	mp4	2	42	1/0
5	Pregrado	Auditivo	Español	mp4	4	9	1/0
6	Pregrado	Visual	Español	jpg	3	71	3/0
7	Pregrado	Visual	Español	jpg	2	36	1/0
8	Educación Superior	Auditivo	Español	mp4	1	71	1/0
9	Básica Secundaria	Visual	Español	mp4	5	71	1/0
10	Carrera Técnica/Tecnológica	Visual	Español	pdf	3	67	1/0
11	Carrera Técnica/Tecnológica	Visual	Español	pdf	3	71	2/1
12	Doctorado	Visual	Español	pdf	2	14	1/0
13	Doctorado	Visual	Español	pdf	2	36	1/0
14	Doctorado	Visual	Español	pdf	2	51	1/0
15	Pregrado	Auditivo	Español	mp4	1	63	1/0
16	Pregrado	Auditivo	Español	mp4	1	67	1/0
17	Doctorado	Visual	Español	pdf	3	63	1/0
18	Pregrado	Visual	Español	jpg	4	63	0/1
19	Pregrado	Visual	Español	jpg	4	67	1/0
20	Pregrado	Visual	Español	jpg	4	71	0/1
21	Carrera Técnica/Tecnológica	Visual	Español	pdf	3	14	1/0

(continued)

Table 7.1 (continued)

Id_case	Level education	Learning style	Language preference	Format preference	Interactivity level preference	Id_LO	Successes/failures
22	Pregrado	Visual	Español	jpg	3	61	1/1
23	Pregrado	Visual	Español	jpg	3	63	2/0
24	Maestría	Visual	Español	mp4	2	14	1/0
25	Educación Superior	Visual	Portugués	pdf	3	67	1/0
26	Educación Superior	Visual	Portugués	pdf	3	71	1/0
27	Pregrado	Visual	Español	pdf	4	9	2/0
…	…	…	…	…	…	…	…

The experiment was performed using the LOs stored in FROAC (http://froac.manizales.unal.edu.co), the Colombian Federation of LORs. Initial searches were performed with Spanish words in order to select the LOs that would initially enter to the recommendation process. In addition, students of Computer/Management Information Systems were selected to use the personal assistant to recommender system, register his/her user profile, and to rank the relevance of the recommendation results.

7.5.1 Study Case

For the study case, we selected a user with these characteristics. Level education: phd, learning style: Visual, language preference: Spanish, format preference: pdf, and interactivity level preference: 2.

The personal assistant performs the verification process to find the similarity of the cases and make the recommendation. Table 7.2 presents the cases and similarity values for this specific user.

The best learning objects or this user are 14, 36, 51, 56, and 63. However, the active user has already rated objects 36, 51 and 63, the result of the recommendation, are the objects with the identifiers: 14, and 56. In Fig. 7.4, we show the interface with the LO recommended. In this interface, the title and description of the LO is presented with the access link; in addition, it presents the number of users similar to those who liked the object (success) and the number of users they did not like the object (failures), this with the aim that the student makes the decision to access it or not.

From this, it is appreciated that the recommendation coincides with LO that in the past were interesting for the user. In the future, it is expected to test with more users and to be able to determine metrics of system visualization.

In the end, we can conclude that the platform includes the proposed model and manages to deliver adaptive learning objects over the student profile. The initial proposal, which aimed at the possibility of using the CBR technique for behavior to intelligence personal assistant, shows its great potential.

Table 7.2 Cases related to study case

Id_case	Level education	Learning style	Language preference	Format preference	Interactivity level preference	Id_LO	Successes/failures	Similarity
28	Doctorado	Visual	Español	pdf	2	14	1/0	1
29	Doctorado	Visual	Español	pdf	2	36	1/0	1
30	Doctorado	Visual	Español	pdf	2	51	1/0	1
31	Doctorado	Visual	Español	pdf	2	56	1/0	1
40	Doctorado	Visual	Español	pdf	3	14	1/0	0,9625
41	Doctorado	Visual	Español	pdf	3	36	1/0	0,9625
42	Doctorado	Visual	Español	pdf	3	56	1/0	0,9625
43	Doctorado	Visual	Español	pdf	3	63	1/0	0,9625
105	Educación Superior	Visual	Español	pdf	2	14	2/0	0,7
106	Educación Superior	Visual	Español	pdf	2	27	1/0	0,7
107	Educación Superior	Visual	Español	pdf	2	51	2/0	0,7
108	Educación Superior	Visual	Español	pdf	2	56	2/0	0,7
109	Educación Superior	Visual	Español	pdf	2	63	2/0	0,7
128	Educación Superior	Visual	Español	pdf	2	36	1/1	0,70
129	Educación Superior	Visual	Español	pdf	2	40	1/0	0,7
130	Educación Superior	Visual	Español	pdf	2	55	1/0	0,7
26	Carrera Técnica/Tecnológica	Visual	Español	pdf	3	67	1/0	0,6625

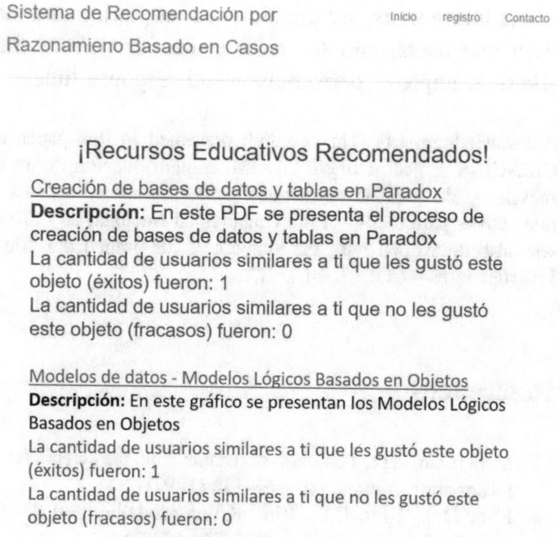

Fig. 7.4 Interface delivering adaptive learning objects

7.6 Conclusions

In this chapter, we propose and demonstrate the possibility of using the CBR technique in the intelligent personal assistant for recommender learning objects, with the aim deliver only adapted materials, and to take advantage of existing cases, shows its great potential. Among the advantages offered by this approach, from the view point of obtaining results for the recommendation and compared with those reported in the prior state of the art, is that our system not requires the explicit rules, not require a complex algorithm, while achieving similar results in using learning objects.

An important factor for this approach to be finalized and reflected in benefits for students is the existence of different educational resources. This work is a sign that it is possible to apply Artificial Intelligence Techniques, particularly in CBR, for the modeling and implementation of recommender systems.

Defining components consistent with the adaptation strategy, that is to say determining the structure of the cases, the way of retrieval, adaptation, evaluation and final storage in the case-base was successful and allowed delivering educational material to students individually.

A rapid satisfaction test was carried out with several users (students of computer systems administration) who were given the recommended objects with the proposed system, and the majority stated that the results were of interest and that the system delivered relevant materials for your search.

This intelligent personal assistant automatically learns from the evaluations of the users because in the attribute success and failure the assistant stores the qualification.

As future work, we are aiming at exploring and incorporating relevant evidence, with real users, and to make evaluations about the learning objects impact. In addition, improve performance and response time.

Acknowledgements The research presented in this paper was partially funded by the COLCIENCIAS project entitled: "RAIM: Implementación de un framework apoyado en tecnologías móviles y de realidad aumentada para entornos educativos ubicuos, adaptativos, accesibles e interactivos para todos" of the Universidad Nacional de Colombia, with code 1119-569-34172. It was also developed with the support of the grant from "Programa Nacional de Formación de Investigadores—COLCIENCIAS".

References

1. Lieberman, H.: Personal assistants for the web: An MIT perspective. In: Intelligent Information Agents, pp. 279–292 (1999)
2. Kim, D.S., Kim, C.S., Rim, K.W.: Modeling and design of intelligent agent system. Int. J. Control Autom. Syst. **1**, 257–261 (2003)
3. Klusch, M., Bergamaschi, S., Petta, P.: European research and development of intelligent information agents: the AgentLink perspective. Intell. Inf. Agents. **2586**, 1–21 (2003)
4. Sikka, R., Dhankhar, A., Rana, C.: A survey paper on E-Learning recommender system. Int. J. Comput. Appl. **47**, 27–30 (2012)
5. Mizhquero, K., Barrera, J.: Análisis, Diseño e Implementación de un Sistema Adaptivo de Recomendación de Información Basado en Mashups. Rev. Tecnológica ESPOL-RTE (2009)
6. Li, J.Z.: Quality, evaluation and recommendation for learning object. Int. Conf. Educ. Inf. Technol. 533–537 (2010)
7. Fleming, N., Baume, D.: Learning styles again: VARKing up the right tree! Educ. Dev. (2006)
8. Alonso, C., Gallego, D., Honey, P.: Los Estilos de Aprendizaje. Procedimientos de diagnostico y mejora, Bilbao (1997)
9. Moreno, R., Duque, N., Tabares, V.: PeCoS-CBR. Personalized courses system with case-based reasoning. IEEExplorer. Computing (2014)
10. Aamodt, A.: Case-Based reasoning: foundational issues. Methodol. Var. Syst. Approach. **7**, 39–59 (1994)
11. Duque, N.: Modelo Adaptativo Multi-Agente para la Planificación y Ejecución de Cursos Virtuales Personalizados Tesis Doctoral (2009)
12. Márquez, V.J.M.: Estado del arte del eLearning. Ideas para la definición de una plataforma universal (2007)
13. Duque, N., Ovalle, D., Moreno, J.: Objetos de Aprendizaje, Repositorios y Federaciones. Conocimiento para Todos (2014)
14. Rossillea, D., Laurentc, J., Burguna, A.: Modelling a Decision-Support System for Oncology using Rule-Based and Case-Based Reasoning Methodologies. Int. J. Med. Inform. **74** (2005)
15. Ali, M., Pan, J., Chen, S., Horng, M.: Mod. Adv. Appl. Intell. (2014)
16. Martinez, N., Garcia, M., Garcia, Z.Z.: Modelo para diseñar sistemas de enseñanza-aprendizaje inteligentes utilizando el razonamiento basado en casos **6** (2009)
17. Lupiani, E., Jimenez, F., Juarez, J.M., Palma, J.: An evolutionary multiobjective constrained optimisation approach for case selection: evaluation in. In: Advances in Artificial Intelligence: 14th Conference of the Spanish Association for Artificial Intelligence, CAEPIA, pp. 396–405 (2014)
18. Costa, C.A., Luciano, M.A.: Information and knowledge models supporting brake friction material manufacturing. J. Brazilian Soc. Mech. Sci. Eng. **26**, 67–73 (2004)

19. Chesani, F.: Recommendation Systems. Corso di laurea Ing. Inform. 1–32 (2007)
20. Burke, R.: Hybrid web recommender systems. Adapt. Web **4321**, 377–408 (2007)
21. Vekariya, V., Kulkarni, G.R.: Hybrid recommender systems: survey and experiments. In: 2012 Second International Conference on Digital Information and Communication Technology and it's Applications (DICTAP), pp. 469–473. IEEE (2012)
22. Burke, R.: Hybrid recommender systems: survey and experiments. User Model. User-adapt. Interact. **12**, 331–370 (2002)
23. Cazella, S.C., Reategui, E.B., Nunes, M.A.: A Ciência da Opinião: Estado da arte em Sistemas de Recomendação, pp. 161–216. JAI Jorn. Atualização em Informática da SBC. Rio Janeiro, RJ PUC Rio (2010)
24. Casali, A., Gerling, V., Deco, C., Bender, C.: Sistema Inteligente para la Recomendación de Objetos de Aprendizaje. Rev. Generación Digit. **9**, 88–95 (2011)
25. Klašnja-Milićević, A., Vesin, B., Ivanović, M., Budimac, Z.: E-Learning personalization based on hybrid recommendation strategy and learning style identification. Comput. Educ. **56**, 885–899 (2011)
26. Salehi, M., Pourzaferani, M., Razavi, S.A.: Hybrid attribute-based recommender system for learning material using genetic algorithm and a multidimensional information model. Egypt. Inform. J. **14**, 67–78 (2013)
27. Zapata, A., Menendez, V., Prieto, M., Romero, C.: A hybrid recommender method for learning objects. Des. Eval. Digit. Content Educ. Proc. Publ. Int. J. Comput. Appl. 1–7 (2011)
28. Klusch, M.: Agent-Mediated trading: intelligent agents and E-Business. Agent Technol. Commun. Infrastruct. 59–76 (2017)
29. Li, F., Sun, J., Zhangi, X.: Study on the key technology of personalized recommendation of case-based reasoning. Int. J. u- e- Serv. Sci. Technol. **8**, 377–382 (2015)
30. da Rocha, A.M., Pereira dos Santos, Jr. V., Michelle, S.: Ambiente de Aprendizagem com Hipermídia Adaptativa. Simpósio Escelência em Gestão e Tecnol. (2012)
31. Alves, P.: Case-based reasoning approach to Adaptive Web-based Educational Systems. IEEExplorer (2008)
32. Ho, W.: Case-based Reasoning: Case Libraries and Analogical Reasoning (2001)
33. Smyth, B.: Case-Based Recomm. Adapt. Web. **4321**, 342–376 (2007)

Part V
Robotics

Chapter 8
Characterize a Human-Robot Interaction: Robot Personal Assistance

Dalila Durães, Javier Bajo and Paulo Novais

Abstract In the last years, the development of robots is entering a new stage where the focus is placed on interaction with people in their daily environments. With the improvement of more and more complex robots to be used in rehabilitation, heath care, service or other applications, robot-human interaction is a rapidly growing area of research. This chapter explores the topic of human-robot interaction. Finally, we presented a proposed framework design that will operate with a person. The system considers the person attitudes level while interact with. The goal is to propose an architecture that monitoring person attitudes in real scenario, and detect patterns of behavior in different occasions. The robot will interact with a person and its training a decision support system that in a real scenario that provide the robot to makes interactions with a person.

8.1 Introduction

The Websters Dictionary defines a robot in three different ways and one of the definitions is "any machine or mechanical device that operates automatically with human-like skill" [1]. Through popular interpretations, these definitions already draw associations between a robot and man.

During the last century robots have operated around humans within industrial and scientific setting. And in the last years, their presence within the home and general

D. Durães (✉) · J. Bajo
Department of Artificial Intelligence, Technical University of Madrid, Madrid, Spain
e-mail: daliladuraes@gmail.com

J. Bajo
e-mail: jbajo@fi.upm.es

D. Durães
CIICESI, ESTGF, Polytechnic Institute of Porto, Felgueiras, Portugal

P. Novais
Departamento de Informática/Centro ALGORITMI, Escola de Engenharia,
Universidade do Minho, Braga, Portugal
e-mail: pjon@di.uminho.pt

A. Costa et al. (eds.), *Personal Assistants: Emerging Computational Technologies*,
Intelligent Systems Reference Library 132, DOI 10.1007/978-3-319-62530-0_8

society becomes ever more common. Frequently robots are used in environments that are inaccessible or unsafe for human beings. Robotic operations include, for example, planetary exploration, search and rescue, activities that impose menacing levels of workload on human operators, and actions requiring complex tactical skills and information integration [20, 27].

The interaction between humans and robots come into physical contact under a variety of circumstances. In the last years, the development of robots is entering a new stage where the focus is placed on interaction with people in their daily environments. The concept of communication with robots has rapidly emerged. The robotics communication will act as a peer providing mental, communicational, and physical support. Such interactive tasks are very importance for allowing robots taking part in human society where many robots have already been applied to various fields in daily environments.

The human-robot interaction is the interdisciplinary study of dynamics interactions between humans and robots. From the point of view of researchers human-robot interactions include a variety of fields, like engineering, computer science, social sciences, and humanities. However, they research subjects are very different: in the field of engineering, the research of study are electrical, mechanical, industrial, and design; in the field of computer sciences, are human-computer interaction, artificial intelligence, robotics, natural language understanding, and computer vision; in the field of social sciences, are psychology, cognitive science, communications, anthropology, and human factor; and in the field of humanities, are ethics and philosophy [13].

From the increasing number of research in this field, it is convenient to distinguish some concepts: such human-robot interaction, social robots, and personal assistance. This chapter deals with the issue of human-robot interaction in personal assistance, with the aim of proving robot, which helps people in their working routines. This chapter is organized as follows. In the next Section the theoretical foundations where scientific literature is reviewed. Section 8.3 contains the proposed design, and finally in Sect. 8.4, discussions and conclusions of this work are presented.

8.2 Theoretical Foundations

Normally, the field of human-robot interaction (HRI) investigate: the development of new techniques for knowledge transfer from human to robot; designing effective tools for human control of a robot; anticipating of the growing presence of robots within general society; and human friendly interface for a robot control.

The goal of HRI is to create teams of humans and robots that are efficient and effective and take advantage of the skills of each team member. An important target of HRI is to increase the number of robotic platforms that can be management by users. For that its necessary have a knowledge of: type of interactions between robots and humans; information that humans and robots need to access, in order to have desirables interchanges; and software architecture that its necessary to accommodate these needs [30].

Table 8.1 A schema of robotics for anthropic domains: main issues and superposition for HRI

Domain	Issues
Design	Lightweight
	Compliance
Control	Safety
	Performance
Sensors	On-line fusion
	Dependability
Biomimetic	Interface
	Human metrics
Software	Open architecture
	Dependability
Planning	Real time
	Consistency

Another target that is keys to the successful of introduction of robots into human environments are safety and dependability. In the field of physical assistance to humans, robots should reduce fatigue and stress; increase human capabilities in terms of force, speed, and precision; and understanding for a correct task execution [10].

In Table 8.1 is presented the fundamentals anthropic domains and the main issues concepts of robotics for anthropic domains for HRI. For each interaction domain humans and robot must have some issues that are important defined. These fundamental domains are design, control, sensors, biomimetic, software and planning. When its planning a HRI interaction, all aspects of design, control, sensors, biomimetic, and software must be considered. Also when the design domain occur, its necessary to considered the planning, the sensors that will be implemented, how the control will be proceed, the biomimetic and the software that will be used. Also when the domain control, software, sensors, and biomimetic will be implemented, its necessary to considered all the other domains. All these domains must be considered together in a HRI, because they are all related with each other. Finally, when its talked about sensors we talked about sensors, actuators, and mechanics, control, and software architectures.

In terms of issues, the fundamental issues are safety, dependability, reliability, failure recovery, and performance. In order to connected all this domains there is a need for pathways connecting crucial components and leading to technological solutions to applications, while fulfilling the viability requirements [10].

In case of robots its necessary considered the design of the mechanism, sensors, actuators, and control architecture in the special perspective for the interaction with humans. In case of humans being its necessary considered control of the mechanism, especially safety and performance, the ways that the interaction occurs in order to following the same metric, and planning. Moreover different roles of interaction with robot are possible, since different people interact in different ways with the same robot, and the robot in turns reacts differently base on its perception of the world.

The interface design is crucial to let the human be aware of the robot possibilities and to provide her/him with a natural way to keep the robot under control at every time.

8.2.1 Social Robots

Dantenhahn and Billard proposed the following definition about social robots: "Social robots are embodied agents that are part of a heterogeneous group: a society of robots or humans. They are able to recognize each other and engage in social interactions, they possess histories (perceive and interpret the world in term of their own experience), and they explicitly communicate with and learn from each other" [9].

In a social robot its important defines some concepts like autonomy, imitation, and privacy.

Autonomy can speed up applications for HRI by not requiring human input, and by providing rich and stimulating interactions. However, autonomy can also lead to undesirable behavior. When a robot has to perform a desired task in a given situation, it is favorable that the constructing system designed has a degree of autonomy. When we talked about sociable robots, its necessary that they have autonomous control in order to interact with humans depending of the situations. Usually, autonomous robots are designed to operate as autonomously and remotely as possible from humans, often performing tasks in dangerous and hostile environments. Other applications such as supplying hospital meals or vacuuming floors bring autonomous robots into environments shared with humans. Although, HRI in these tasks still minimal.

Imitation occurs when robot intended to imitate the human being. However, negative correlation between the robots physical realism and its effectiveness in HRI can happen when physical similarity that attempts in imitation of human-like appearance and behavior could cause discord.

Privacy is a question that is presented when the presence of a robot inherently affects a users sense of privacy [21]. Because of its synthetic nature, a robot is perceived as less of a privacy invasion than a person, especially in potentially embarrassing situations.

Social robot can be interpreted as the interface between man and technology. However its considered socially interaction robots that exhibit the following characteristics: express and/or perceive emotions; communicate with high-level dialogue; learn/recognize models of other agents; establish/maintain social relationships; use natural cues (gaze, gestures, etc.); exhibit distinctive personality and character; and may learn/develop social competencies. This type of robots can be used for variety purposes: as educational tools, as therapeutics aids, or as toys.

In social interaction robots can operate as an assistants, peers, or partner, which imply that they needs to have a certain degree of flexibility and adaptability, in order to interact with humans. Robots that are socially interactive can have different forms and functions, ranging from robots whose only purpose is to have a single task to robots that have a collection of tasks [6]. It is the use of social acceptable interaction between robots and humans that helps break down the barrier between the digital information space and the human being. These interactions may characterize the first stages where people stop perceiving machines as simply tools.

Social robots interactions are important in a wide range of domains. One example is the interaction where robots need to exhibit peer-to-peer interaction skills. In this case its necessary that robot solve specific tasks and interact socially with humans.

If we consider the situation where robot accompanies elderly care at home. In this case the robot may improve skills in order to maintain the elderly people interest. In these situations it may be desirable for a robot that he develops its interactions skills over the time.

The degree of social robot interaction is accomplished through a progressive and adaptive process. Its possible to considerer a minimum requirement for social robots interactions, which is the ability to adapt to social situations and understanding and communicate with.

8.2.2 *Personal Assistance*

In the last years, especially in domestic, entertainment, and health care a new range of application domains has emerged where robots can interact and cooperate with humans as a partner or as a peer.

Humans learn through a range of techniques including observation, imitation, instruction, and simulation [14].

An individual interacts with his social environment to acquire new competencies. With social robots its necessary that they learning with the environment that they interact. The problem of learning is that the robot needs to distinguish the correct actions and state in order to create new policy that enables a robot to select an action based upon its current world state. Additionally, because of differences in sensing and perception, robot may have very different views of the world. Thus, learning is often essential for improving communication, facilitating interaction, and sharing knowledge [23]. The way that the robot might learn can be very different. When its addressed to robot-robot work, their communication can be the "leader following" [8, 19], inter-personal communication [3, 5, 31], imitation [4, 15], and multi-robot formations [25].

In case of HRI there exits some approach to learning. One is to create sequences of known behaviors in order to match human models [24]. Another is to match observations to known behavioral such as motor primitives [11, 12]. Finally, the most common is imitation.

An intelligent personal assistant is an application or a robot that uses inputs such as users voice, vision (images), and contextual information to provide assistance by answering questions in natural language, making recommendations, and performing actions.

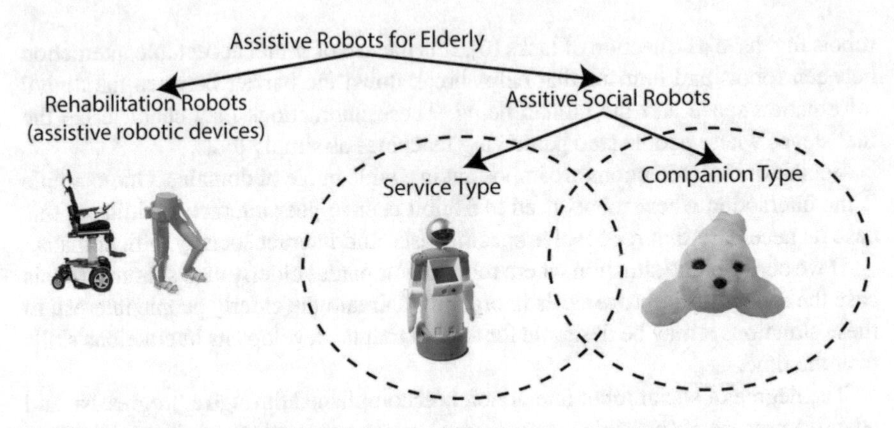

Fig. 8.1 Categorization of assistive robots for elderly

Nowadays, platforms with ambient intelligent and robotics are developing quickly and the results are products that have the potential to play an important role in assisting the elderly [28]. In health-care its required to have robust information with respect to their effects, which is necessary to used technology in an effective and efficient way.

In elderly care, personal assistant can have two types of robots: rehabilitation robots or social robots, which in Fig. 8.1 are presented these two types of robots.

The first type of robots its can use physical assistive technology that is not primarily communicative and isn't destined to seem as a social entity. Examples of this type of technology are exoskeleton [22], artificial limbs, and smart wheelchairs [16].

The second type of robots concerns systems that can be perceived as social entities that communicate with the user. Examples of this type of robots are service type and companion type.

The service type robots are robots that are used as assistive devices. The social functions of such service type robots exist mostly to facilitate interfacing with the robot. There functionalities are related to the maintenance of independent living by supporting basic activities and mobility. The basics functionalities are eating, bathing, toileting and get dressed. The functionalities related with mobility include navigation and provide household maintenance, monitoring of those who need continuous attention and preserving safety [2].

Companion type robot is study in the companionship that a robot might provide. The main functionalities of these robots are to enhance healthiness and psychological pleasure of elderly users by providing companionship. Social functions implemented in companion robots are principally aimed at growing well-being and psychological happiness.

Nevertheless, there are robots that have the two functions: they can be companion robots as well service robot.

8.2.2.1 Personalization

There is not a single accepted definition about the meaning of personalization. So we can define personalization in very different ways:

- As a system with methods that incorporate technology in order to differ from resources and processes, based on each learners skills, interests, and needs and learning profile in order to accelerate and deepen learning [17, 32].
- Personalized service mentions to any behavior happening in a service interaction intended to individuate customer and service experience [33].
- Personalization is the ability to supply services and actions that can responds to the users requirements and goals based on deep knowledge about personal preferences and behavioral obtained through client monitoring [29].
- Personalization is the result of intimate relationship and knowledge about a user. When a relationship with a user increases the level of personalization can also increase. Personalization is intended to facilitate a process of interactions between the robots and the users.

A robot that remembers and recognizes its past interactions with users might give them the feeling of getting special attention and personal recognition when they meet the robot again. The feeling of being treated as special is one of the reasons why users build relationships [18].

Recommendation systems are a main point of research in relation to personalization, and different variables are used to control the recommendations: learning styles, performance, learners activities, browsing behaviors, learners interests or social connections among others. This type of systems can better predict and anticipate the needs of users, and act more efficiently in response to their behavior [35].

8.3 The Proposed Design

In our western population there is a growing necessity for new technologies that can assist and care the elderly in their daily lives routines. There are two reasons for this. First, people prefer more and more to live in their own homes as long as possible instead of being institutionalized in sheltered homes, or nursery homes when problems related to ageing appear. Second, it is expected that western countries will face a tremendous shortage on staff and qualified healthcare personnel in the near future [26].

The quality of life for people remaining in their own homes is generally better than for those who are institutionalized. Furthermore, the cost for institutional care can be much higher than the cost of care for a patient at home. To balance this situation, efforts must be made to move the services and care available in institutions to the home environment.

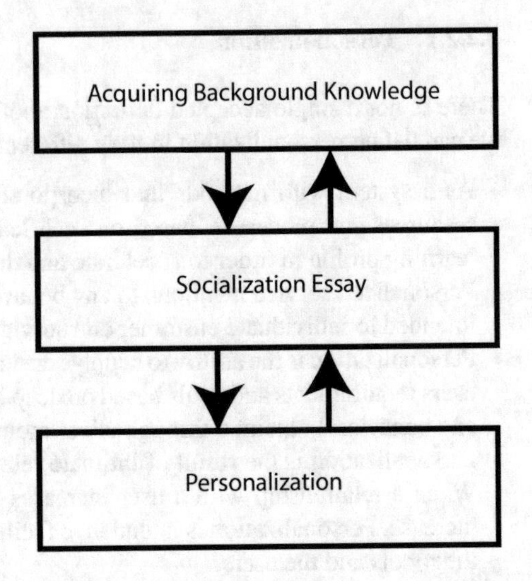

Fig. 8.2 General model for a personal assistant to elderly or persons with mobility impairments

However, human nature created persons that are individual and personal characteristics. Elderly people and people with mobility impairments have individual needs and specific characteristics that a companion would have to adapt to.

For this reason its proposed to develop a personalized robot that can serve as companion and that can adapt to the needs and interaction styles of elderly or those with mobility impairments that they are interact with. Such robot will be personalized, which individually reflects the needs and requirement of the social environment where the robot is operating in.

In Fig. 8.2 and Table 8.2 its presented a general preview model for a personal assistant to elderly or person with mobility impairments.

In the first phase its necessary to have extensive user studies need, as well as appearances influence peoples attitudes, opinions and preferences towards robots. Its also required knowing the tasks that the robot is supposed to perform, the physical environment that the robot is operating in, as well as the social environment.

In the second phase its the socialization essay. Based on the generic knowledge acquire in phase one, a first prototype of the robot can be tested and redefined in controlled environmental conditions, in order to determine the default settings. In this phase, basic behavioral patterns for the robot are defined and personality will be formed. The information about personality profiles as well as requirements and constraints derived from the tasks or environment that the robot is supposed to perform are also defined.

Finally, in phase three the robot will be personalized. In this phase personality profile and other information that can be acquire about people and environment that the robot interact will be adjust on robot behavior repertoire. Once the robot is place in home he will interact with person that it is supposed to live with. The robot needs to adjust their default settings and learn from their experience.

Table 8.2 Model for tasks to be carried out in each phase of HRI

Phase	Task to be carried out
Acquiring background knowledge	Defining tasks to robot perform
	Knowing physical environment
	Knowing social environment
	Defining user profiles
Socialization essay	Defining a prototype of the robot
	Defining basic behaviors patterns
	Defining reactiveness /autonomous robot
Personalization	Adjustment robot behavior
	Interactions histories
	Learning from experience
	Adaption on social learning

Related to the social behavior that influences the HRI various parameters need to be identified in HRI studies. Examples of these parameters are interaction distance, seeking attention, reactiveness/autonomy, and expression of intentionality. All of these parameters might be different and depends on the environment and the profile of the user that the robot will interact.

As a result of this process even two robots with the same structured and initial defining settings, will over the time develop individualized settings creating a unique personality. Such robot will have to be able to manage with changes in relationship with elderly people and persons with mobility impairments.

Figure 8.3 depicts the process through which the system operates; it is possible to observe the different classifications of information in order to allow, in the end, the management of HRI.

8.3.1 Dynamic HRI Monitoring Architecture

The robot must have a platform that allows moving in every direction. Consequently the robot should be omnidirectional, with three wheels. The wheels are placed at 120 between them. The robot should be used a Arduino microcontroller and be constructed with a several separate modules in order to be easily changed, which makes it possible to changed robot functions quickly and safely. The height of the robot will be around 60 cm, which is the ideal height for the robot to interact with people sitting on a chair or lying on a bed. When the battery drops below 10%, the robot will autonomously move to the doc-station, where it will charge.

A survey will collect information about the most common needs of the person. This module, upon converting the sensory information into useful data, allows for a contextualized analysis of the operational data of the persons actions and this frame-

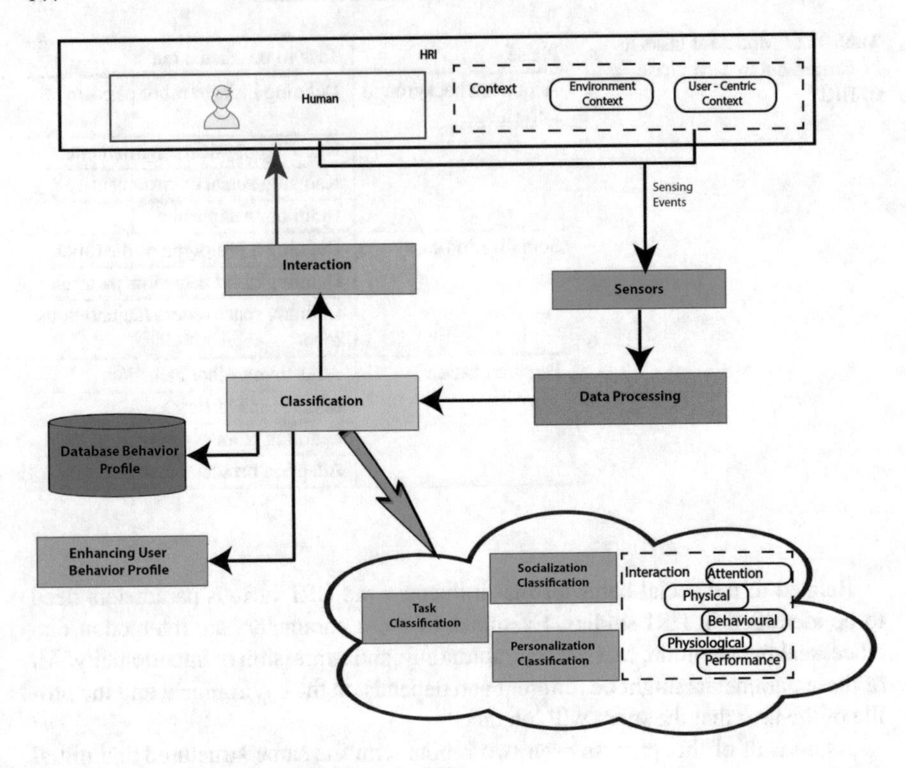

Fig. 8.3 Dynamic interface architecture for HRI

work performs this contextualized analysis. Then, the persons profile is updated with new data, and the robot acts in order to the feedback from this module. The system is developed to acquire data from normal working compiles information from persons activities.

The proposed framework includes not only the complete acquisition and classification of the data, but also an interaction level that will support the human-based or autonomous decision-making mechanisms that are now being implemented.

The Sensing Events are charged for capturing information describing the behavioral patterns of the persons, and receiving data from context environment. This layer encodes each event with the corresponding necessary information. These data are further processed, stored and then used to calculate the values of the behavioral person.

The Data Processing layer is responsible to process the data received from the Sensor layer in order to be evaluate those data according to the metrics presented. Its important that in this process some values should be filtered to eliminate possible negative effects on the analysis. The system receives this information in real-time and calculates, at regular intervals, their position and the interaction that must occurred.

The Classification layer is where the indicators are interpreted for example: interpreting data from the interaction indicators and to build the meta data that will support decision-making. When the system has an enough large dataset that allows making classifications with precision, it will classify the inputs received into different interaction levels in real-time. This layer has access to the current and historical state of the group from a global perspective, but can also refer to each person.

For that, this layer uses the machine learning mechanisms. After the classification, the Enhancing User Behavior Profile layer is responsible for providing access to the lower layer. The Database Behavior Profile is also a very important aspect to have control off. This possibility allows to analyses within longer time frames. This layer, whose function detect persons mood preserving those information (actual and past) in the mood database. This information will be used by another sub-module, the affective adaptive agent, to provide relevant information to the platform and to the mentioned personalization module.

Finally at the top, the Interaction layer includes the mechanisms to build intuitive actions, language and visual representations that make the robot interact with a person. At this point, the system can start to be used by the people involved, especially a supervisor who can better adapt and personalize his strategies. The actual persons mood information is used in the Interaction layer, and can be used to personalize actions according to the specific person.

8.4 Discussions and Conclusions

In this chapter its presented a framework for a social robot in order to act as a person, and more especially for personal assistance for elderly people and people with mobility impairments. This robot will possess individual and social learning skills, which make it unique.

Predicting preferences of elderly and persons with mobility impairments and providing the personalized robots based on persons preferences are important issues. However, the research for offering robot personalized considering the persons preference on context-aware computing is a relatively insufficient research field.

Nevertheless the robots are developed and deployed for the purpose of solving social problems, however its impact on the lives of the elderly and people with mobility impairments, there are social and ethical implications associated with the deployment of the robots. The advantages associated with the use of social robots in healthcare settings are largely dependent on the process of personalization, in order to facilitate the human-robot relationships [7].

Although, the ability for robots to interact with people and to control from these interactions to perform tasks better, to promote their self-maintenance, and to learn in an environment as complex as that of humans is of tremendous pragmatic and functional importance for the robot.

Suitable personalization is necessary to meet peoples needs and to ensure that robots could function independently to respond to people and unfamiliar situations. But it also raises ethical and social concerns, such as the tension between personalization, safety, and privacy [34].

When designing process of personalization, collection of personal data is necessary for the social robot to be personalized to meet the purpose for which it was designed.

For the point of view of robots, they not only have to carry out their tasks, but also have to survive in the human environment. From the robots perspective, the real world is complex, unpredictable, partially knowable, and continually changing. The capability of robots to adapt and learn in such an environment is essential.

This research suggests the basic direction for provision of the personalized services of robot and utilization of context history. Additionally, this research can be the basic direction of design and the guidelines of development for personalized robots. However, the prototype was not implemented according to the proposed framework. Also, the protection of personal information or privacy needs to be considered.

Acknowledgements This work has been supported by COMPETE: POCI-01-0145- FEDER-007 043 and FCT Fundao para a Cincia e Tecnologia within the Project Scope: UID/CEC/ 00319/2013.

References

1. Websters dictionary. http://www.dictionary.com/browse/robot?s=t
2. Bahadori, S., Cesta, A., Grisetti, G., Iocchi, L., Leone, R., Nardi, D., Oddi, A., Pecora, F., Rasconi, R.: Robocare: an integrated robotic system for the domestic care of the elderly. In: Proceedings of Workshop on Ambient Intelligence AI* IA-03, Pisa, Italy. Citeseer (2003)
3. Billard, A., Dautenhahn, K.: Grounding communication in autonomous robots: an experimental study. Robot. Auton. Syst. **24**(1–2), 71–79 (1998)
4. Billard, A., Dautenhahn, K.: Experiments in learning by imitation-grounding and use of communication in robotic agents. Adapt. Behav. **7**(3–4), 415–438 (1999)
5. Billard, A., Hayes, G.: Learning to communicate through imitation in autonomous robots. In: International Conference on Artificial Neural Networks, pp. 763–768. Springer (1997)
6. Breazeal, C.L.: Designing Sociable Robots (Intelligent Robotics and Autonomous Agents). MIT Press, Cambridge (2002)
7. Dahl, T.S., Boulos, M.N.K.: Robots in health and social care: a complementary technology to home care and telehealthcare? Robotics **3**(1), 1–21 (2013)
8. Dautenhahn, K.: Getting to know each other artificial social intelligence for autonomous robots. Robot. Auton. Syst. **16**(2–4), 333–356 (1995)
9. Dautenhahn, K., Billard, A.: Bringing up robots or the psychology of socially intelligent robots: from theory to implementation. In: Proceedings of the Third Annual Conference on Autonomous Agents, pp. 366–367. ACM (1999)
10. De Santis, A., Siciliano, B., De Luca, A., Bicchi, A.: An atlas of physical human-robot interaction. Mech. Mach. Theory **43**(3), 253–270 (2008)
11. Demiris, J., Hayes, G.: Imitative Learning Mechanisms in Robots and Humans. University of Edinburgh, Department of Artificial Intelligence (1996)
12. Demiris, J., Hayes, G.: Active and passive routes to imitation. In: Proceedings of the AISB99 Symposium on Imitation in Animals and Artifacts, pp. 81–87 (1999)
13. Feil-Seifer, D., Matarić, M.: Using proxemics to evaluate human-robot interaction. In: Proceedings of the 5th ACM/IEEE International Conference on Human-robot Interaction, HRI '10, pp. 143–144. IEEE Press, Piscataway, NJ, USA (2010). http://dl.acm.org/citation.cfm?id=1734454.1734514

14. Galef, J., Bennett, G.: Imitation in animals: history, definition, and interpretation of data from the psychological laboratory. In: Social Learning: Psychological and Biological Perspectives, p. 28 (1988)
15. Gaussier, P., Moga, S., Quoy, M., Banquet, J.P.: From perception-action loops to imitation processes: a bottom-up approach of learning by imitation. Appl. Artif. Intell. **12**(7–8), 701–727 (1998)
16. Gomi, T., Griffith, A.: Developing intelligent wheelchairs for the handicapped. In: Assistive Technology and Artificial Intelligence, pp. 150–178 (1998)
17. Grant, P., Basye, D.: Personalized Learning. International Society for Technology in Education, Arlington (2014)
18. Gwinner, K.P., Gremler, D.D., Bitner, M.J.: Relational benefits in services industries: the customers perspective. J. Acad. Mark. Sci. **26**(2), 101–114 (1998)
19. Hayes, G.M., Demiris, J.: A Robot Controller Using Learning by Imitation. University of Edinburgh, Department of Artificial Intelligence (1994)
20. Hinds, P.J., Roberts, T.L., Jones, H.: Whose job is it anyway? a study of human-robot interaction in a collaborative task. Hum. Comput. Interact. **19**(1), 151–181 (2004)
21. Kahn, P.H., Ishiguro, H., Friedman, B., Kanda, T.: What is a human?-toward psychological benchmarks in the field of human-robot interaction. In: The 15th IEEE International Symposium on Robot and Human Interactive Communication, 2006 (ROMAN 2006), pp. 364–371. IEEE (2006)
22. Kazerooni, H.: Exoskeletons for human power augmentation. In: 2005 IEEE/RSJ International Conference on Intelligent Robots and Systems, 2005 (IROS 2005), pp. 3459–3464. IEEE (2005)
23. Klingspor, V., Demiris, J., Kaiser, M.: Human-robot communication and machine learning. Appl. Artif. Intell. **11**(7), 719–746 (1997)
24. Mataric, M.J., Williamson, M., Demiris, J., Mohan, A.: Behavior-based primitives for articulated control. In: Fifth International Conference on Simulation of Adaptive Behavior (SAB-98), pp. 165–170 (1998)
25. Michaud, F., Letourneau, D., Guilbert, M., Valin, J.M.: Dynamic robot formations using directional visual perception. In: IEEE/RSJ International Conference on Intelligent Robots and Systems, 2002, vol. 3, pp. 2740–2745. IEEE (2002)
26. Miskelly, F.G.: Assistive technology in elderly care. Age Ageing **30**(6), 455–458 (2001)
27. Parasuraman, R., Cosenzo, K.A., De Visser, E.: Adaptive automation for human supervision of multiple uninhabited vehicles: effects on change detection, situation awareness, and mental workload. Mil. Psychol. **21**(2), 270 (2009)
28. Pollack, M.E.: Intelligent technology for an aging population: the use of ai to assist elders with cognitive impairment. AI Mag. **26**(2), 9 (2005)
29. Riecken, D.: Personalized views of personalization. Commun. ACM **43**(8), 26–26 (2000)
30. Scholtz, J.: Theory and evaluation of human robot interactions. In: Proceedings of the 36th Annual Hawaii International Conference on System Sciences, 2003, 10 pp. IEEE (2003)
31. Steels, L.: Emergent adaptive lexicons. Anim. Animat. **4**, 562–567 (1996)
32. Steiner, E.D., Hamilton, L.S., Peet, E., Pane, J.F.: Continued Progress: Promising Evidence on Personalized Learning (2015)
33. Surprenant, C.F., Solomon, M.R.: Predictability and personalization in the service encounter. J. Mark. 86–96 (1987)
34. Sutanto, J., Palme, E., Tan, C.H., Phang, C.W.: Addressing the personalization-privacy paradox: an empirical assessment from a field experiment on smartphone users. MIS Q. **37**(4), 1141–1164 (2013)
35. Verbert, K., Manouselis, N., Ochoa, X., Wolpers, M., Drachsler, H., Bosnic, I., Duval, E.: Context-aware recommender systems for learning: a survey and future challenges. IEEE Trans. Learn. Technol. **5**(4), 318–335 (2012)

Chapter 9
Collaboration Between a Physical Robot and a Virtual Human Through a Unified Platform for Personal Assistance to Humans

S.M. Mizanoor Rahman

Abstract Two human-like intelligent artificial characters (agents) of heterogeneous realities (a humanoid robot and a virtual human) are developed with similar intelligence and functionalities. Operations of the characters are integrated through a unified platform. The characters are separately deployed in a homely environment to provide personal assistance to a physically disabled human in finding a missing household object. Human's trust in the characters as well as bilateral trust between the characters for the collaboration are separately modeled, and real-time trust measurement methods are developed. A collaboration scheme between the characters based on the bilateral trust is developed. A comprehensive evaluation scheme is developed to evaluate the performance of the characters in assisting the human. The performance levels of the individual characters and their collaboration toward providing personal assistance to the human are compared. The results show that the individual characters and their collaboration can assist the disabled human successfully, but with varying capabilities. For example, the collaboration partly outperforms the individual characters, and the physical character (robot) partly outperforms the virtual character. The results are novel that broaden the horizon of humanoid robots and virtual humans and open a new paradigm of collaboration between physical (robot) and virtual (virtual human) characters. The results are useful to develop artificial characters of heterogeneous realities and their social collaborations to provide personal assistance to disabled humans in their daily living that may improve the quality of life and produce positive economic and societal impacts.

Keywords Assisted living · Bilateral trust · Personal assistance · Robot-virtual human collaboration · Unified platform

---

S.M. Mizanoor Rahman (✉)
Department of Mechanical Engineering, College of Engineering and Sciences, Clemson University, Clemson, SC 29634, USA
e-mail: rsmmizanoor@gmail.com

A. Costa et al. (eds.), *Personal Assistants: Emerging Computational Technologies*, Intelligent Systems Reference Library 132, DOI 10.1007/978-3-319-62530-0_9

9.1 Introduction

A Personal Assistant (PA) is an agent capable of assisting humans performing tasks in their workplaces and daily activities [1, 2]. Personal assistants may be used for accessing information from databases autonomously to guide humans through different tasks and deploying learning mechanisms to acquire new information on human user's performance [2]. Currently, personalization is gaining more and more attention and priority [3]. However, the state-of-the-art personal assistants in most cases are software-generated special applications [1, 3]. Intelligent mobile personal agents such as mobile humanoid robots that can co-exist with humans in physical environment and assist humans in their daily living are very rare [3]. It is assumed that the use of humanoid robots as personal assistants to humans in homely environment has not still reached its maturity due to lack of required intelligence, autonomy, functionalities, human-friendly interaction abilities, alignment with practical application scenarios, etc. It is assumed that home-based settings may be less expensive and more readily accessible to humans over institutional settings. The state-of-the-art robots as personal assistants are also neither so personalized nor very much suitable for home-based daily assistance [4–8]. In addition, the state-of-the-art humanoid robots are not sufficiently enriched with ambient intelligence [9]. They are also not so natural, human-like, and cheap [10]. The robots are not enriched with adequate softwares, and the used technologies are not so advanced. For example, the emerging technologies such as agent ecology [9], cyber-physical system [11] and IoT [2] are not sufficiently integrated with robotics technologies, and the robots also do not possess advanced learning and adapting capabilities [12].

On the other hand, virtual humans are software-generated human-like animated artificial agents [13]. Present applications of virtual humans include various tasks such as serving as virtual patient, tutor, student, trainee, advertiser, and so forth [13–15].There are increasing contributions of virtual humans toward anatomy education, psychotherapy and biological and biomedical research [16, 17]. However, it appears that virtual humans could not come beyond their virtual environments, i.e. their applications are limited to the virtual world, and their interactions with their real-world human counterparts are still limited [18]. It is believed that the scope of their contributions could be augmented if they could be used to assist humans or cooperate with humans to perform real-world tasks [19]. The virtual humans may be enriched with various real-world functions and attributes for interactions with their human counterparts such as they may exhibit human-like intelligence, motions, actions, emotions, gestures and expressions, communicate and interact with humans, memorize facts and retrieve according to dynamic contexts, and demonstrate reasoning and decision-making abilities based on their perceptions [20]. The virtual humans may be less costly and less affected by environmental disturbances and constraints compared to robots, which may make them suitable as low-cost personal assistants in homely environments [21, 22].

However, such real-world assistance of virtual humans to humans is still not a priority.

A robot and a virtual human have a lot in common in their objectives and performance. Robots and virtual humans may separately assist humans and also collaborate with each other to assist humans performing various real-world tasks [18, 19]. Networked and bi-laterally communicating robots and virtual humans cooperating in a coordinated and goal-oriented way may assist humans in better ways than an individual robot or a virtual human [18, 19]. Dynamic collaboration between robots and virtual humans seems to be superior to the augmented reality for robots where a robot may follow its virtual counterpart, but dynamic bidirectional collaboration between them is limited [23]. A comprehensive framework is necessary to stage real-world collaboration between a robot and a virtual human. However, investigations on collaborations between robots and virtual humans have not received much attention yet except a few preliminary initiatives that are in concept design phases, and no real agents and cooperation methods have been proposed to justify the effectiveness of such initiatives [24, 25]. Again, a common platform may be helpful to implement the real-world collaboration between a robot and a virtual human, e.g. it may reduce the volume of software development and ease the animation. However, suitable framework for collaboration between a robot and a virtual human and an initiative to develop a common platform between them have not received much importance.

Well-defined comprehensive evaluation schemes are necessary to evaluate the performance of the assistant agents, which may increase their social acceptance and impacts [26, 27]. However, suitable evaluation schemes for personal assistant agents are also not available. Human's trust in personal assistants and trust of one assistant agent in another agent are mandatory for any assistive task for humans [28]. Human trust in collaborating robots has been studied enormously [29], but human trust in collaborating virtual humans as well as trust between two artificial agents of heterogeneous realities (e.g., robot's trust in virtual human and virtual human's trust in robot) have not been studied yet. The trust in each other may be used to plan their role and autonomy in their collaboration. Appropriate computational models of trust are necessary to measure the real-time trust of human in robot and virtual human and the trust between the robot and the virtual human for their collaboration. However, such trust modeling and real-time trust measurement methods have not been proposed yet. On the other hand, mixed-initiative is the case where the turns of leaderships in collaboration are negotiated between participating agents rather than solely determined by a single agent [30]. It is assumed that mixed-initiatives between a robot and a virtual human may make their collaboration more participatory, intuitive and natural, which may enhance their individual contribution to the collaboration toward assistance to humans. Bilateral trust status between a robot and a virtual human may trigger their taking turn in the mixed-initiative collaboration. However, bilateral trust-triggered mixed-initiatives in collaboration between a robot and a virtual human for assistance to humans has not been investigated yet.

Being motivated by the above facts and figures, the objective of this chapter is determined as to develop an intelligent physical agent (a robot) and a virtual human as personal assistants to disabled humans in their homely environment. The agents assist the humans individually and collaboratively being unified through a common platform. The assisted human's trust in the assistant agents as well as trust between the artificial assistant agents are modeled and measured. Individual assistance and assistance through trust-based collaboration are evaluated and compared.

Organization of the chapter is as follows: Sect. 9.2 highlights the related works. Section 9.3 presents the development of the personal assistant robot and the virtual human. Section 9.4 presents the unified platform to integrate the operations of the robot and the virtual human. Section 9.5 demonstrates home-based settings to assist disabled persons in their daily living by the robot, the virtual human and their bilateral trust-based collaboration. Section 9.6 illustrates the modeling and measurement of human trust in robot and virtual human and bilateral trust between the robot and the virtual human. Section 9.7 introduces the evaluation scheme to evaluate the assistance of the robot, virtual human and their collaboration to the disabled humans. Section 9.8 presents the experimental evaluation of the quality of the assistance of the robot, virtual human and their collaboration to the disabled humans. Section 9.9 discusses the limitations of the presented methods and results. Section 9.10 draws conclusions and shows the future directions of the research. Then, the acknowledgement and the references are presented.

9.2 Related Works

Different types of personal assistants have been developed with different objectives in recent years. Chen and Barthes proposed a memory mechanism for personal assistants in order to enhance agent intelligence while working with the user or other agents [1]. Santos, Rodrigues, Casal, Saleem and Denisov proposed the integration of personal assistants into ubiquitous computing environments in an Internet of Things (IoT) context considering many different factors such as heterogeneity of objects and diversity of communication protocols and enabling technologies [2]. Homayounvala, Aghvami and Groves proposed personal assistant agents to observe user behavior and make mobile services more valuable for users by making them easier to use and more adaptive [3]. Jones, Moulin, Barthes, Lenne, Kendira and Gidel proposed voice-controlled personal assistant agents to be implemented to provide unique interactions within a multi-surface environment [31]. Wahaishi and Aburukba presented an agent-based personal assistant architecture that provides an innovative approach to automate the exam scheduling processes allocating different proctors and exam rooms [32]. Sugawara, Manabe and Fujita proposed a concept of a mobile symbiotic interaction between a user and a personal assistant to watch over the particular user using sensors deployed in a ubiquitous environment [33].

Czibula, Guran, Czibula and Cojocar proposed a personal assistant agent that learns by supervision to assist users in performing specific tasks [34]. Wong, Aghvami and Wolak developed context-aware personal assistant agent-based systems and concepts to provide computing environments [35]. Bush, Irvine and Dunlop proposed a user-centric system that aims to hide complexity from the user incorporating concepts of the personal distributed environment [36]. Blake discussed the rationale for a personal learning assistant agent customized for business processes [37]. Ma, Feng, Yang and Wu proposed an agent-based personal article citation assistant [38]. Nack, Roor, Karg, Kirsch, Birth, Leibe and Strassberger proposed a personal assistant to provide mobility models for travels and transportation [39]. Jalaliniya and Pederson proposed wearable personal assistants for surgeons in healthcare [40], etc.

Sansen, Torres, Chollet, Glackin, Delacretaz, Boudy, Badii and Schlogl proposed a personal assistant vocal humanoid robot with verbal and non-verbal communication capabilities that can provide care to dependent persons at home [41]. Koubaa, Sriti, Javed, Alajlan, Qureshi, Ellouze and Mahmoud proposed the design of an assistive mobile robot to support people in their everyday activities in office and home environments [4]. Qiu, Ji, Noyvirt, Soroka, Setchi, Pham, Xu, Shivarov, Pigini, Arbeiter, Weisshardt, Graf, Mast, Blasi, Facal, Rooker, Lopez, Li, Liu, Kronreif and Smrz proposed personal assistant robots that are robust [5]. Mishra, Makula, Kumar, Karan and Mittal proposed a voice-controlled personal assistant robot for applications at homes, hospitals and industries [6]. Luria, Hoffman, Megidish, Zuckerman and Park proposed a personal assistant robot for use at homes [7]. Webster, Dixon, Fisher, Salem, Saunders, Koay, Dautenhahn and Saez-Pons proposed a high-level planner/scheduler for the care-o-bot, an autonomous personal robotic assistant [8], etc. In addition, Arafa and Mamdani [21], and Matsuyama, Bhardwaj, Zhao, Romero, Akoju and Cassell [22] proposed virtual personal assistants though these are still not robust to provide assistance to humans at homes.

9.3 Development of the Personal Assistant Robot and the Virtual Human

The hardware components, software packages, control/operation methodologies and communication technologies for the humanoid robot and the virtual human are determined, and the agents are developed using these facilities with appropriate functions and intelligence so that the agents are able to perform as personal assistants to humans for the selected task. Development of the agents is described below.

9.3.1 Development of the Humanoid Robot

A NAO humanoid robot (http://www.aldebaran-robotics.com/en/) as shown in Fig. 9.1a is properly developed and used as an intelligent personal assistant to disabled humans. For this purpose, the application programming interfaces (APIs) for various functions such as wave hand, stand up, sit down, shake hand, walk, point at something, grab and release an object, speech (text to speech), look at a position, etc. are developed. The APIs for the functions and the control softwares are archived in a robot control server. The robot is also made able to perceive the environment through sensors (e.g., vision cameras), make decision based on adaptive rules and stored information, and react by talking, moving or showing emotions. Thus, the functions along with sensory information make the robot intelligent and skillful for assisting the human in the context of a selected representative task.

9.3.2 Development of the Virtual Human

A virtual human is developed as shown in Fig. 9.1b using the Smartbody system (http://smartbody.ict.usc.edu/) for its animation and control. The virtual human model is developed based on the joints and skeleton requirements of the smartbody, and it is exported to the 3D Autodesk Maya system (http://www.autodesk.com/). The specifications of the virtual human's physical configuration follow biomimetics approach [10], i.e., the hand movement and locomotion speed, trajectories for gestures, body dimensions, joint angles, etc. are determined being inspired by these for humans. The Ogre system (http://www.ogre3d.org/) is used for graphical

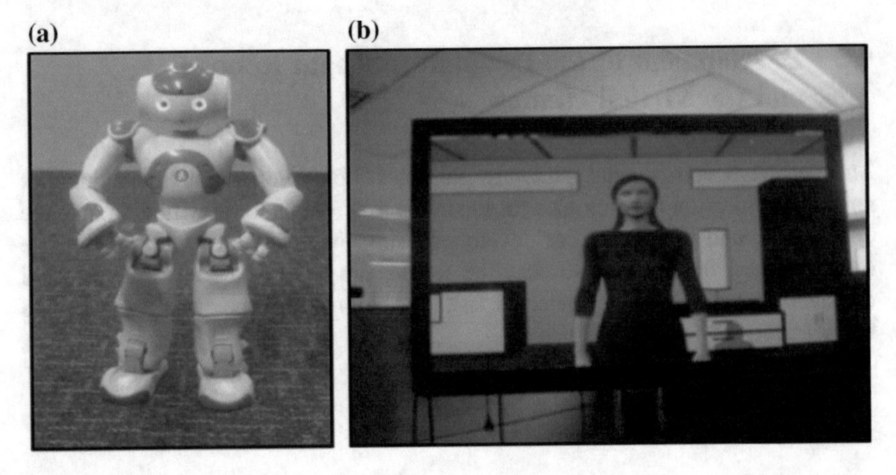

Fig. 9.1 The intelligent personal assistant agents, **a** a humanoid robot, **b** a virtual human

rendering. APIs for various functions such as action, locomotion, recognition, gesture, expression, etc. are developed and archived in the control server for the virtual human. The functions include various gestures (e.g., turn head, look at a position), text to speech conversion, gaze, object manipulation, locomotion (walk to a position), different facial expressions (emotions), actions, etc. The virtual human is then displayed in a screen as shown in Fig. 9.1b.

9.4 The Unified Platform to Integrate the Operations of the Robot and the Virtual Human

A common communication platform is developed to integrate the virtual human and the humanoid robot, and then installed in a computer system as shown in Fig. 9.2. The common platform works in such a way that animation of each function for the robot or the virtual human can be commanded from a common client, which is networked with the control server through a *Thrift* interface [42]. A remote procedural call (RPC) library is used to handle the communication between the command script and the control server [43]. RPC is modular and flexible, and it relies on a server/client relation allowing inter-process communication, which is the motivation behind using the RPC. Thrift is preferred over robot operating system (ROS) because the ROS runs only on Linux/CORBA, and it is complex [44]. Instead, Thrift is reliable, it supports cross-platform/cross-language, and provides good performance. The virtual human control server is connected to a display window within the computer and the virtual human is also displayed in an external display screen. The humanoid robot control server is connected to the physical robot using appropriate wireless network. This novel unified platform is named as "*Common Communication Platform (CCP)*" for heterogeneous agents. The common platform can be used to operate agents of heterogeneous realities such as a

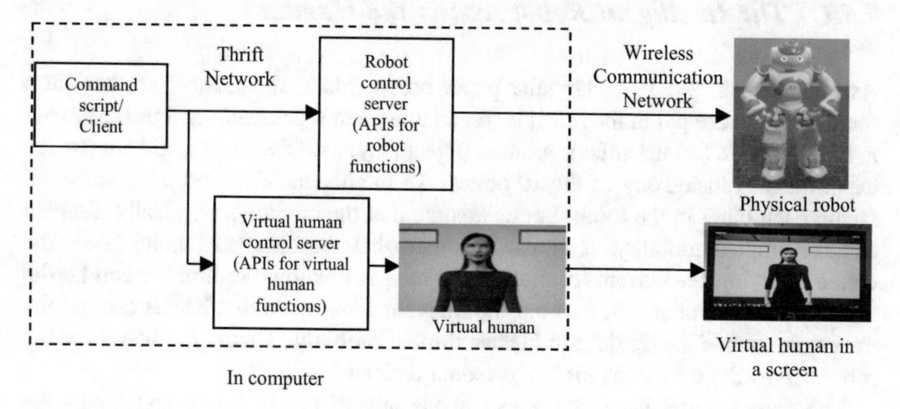

Fig. 9.2 The unified/common communication platform (CCP) for the virtual human and the humanoid robot

physical robot and a virtual human specifying the agent's name (e.g., robot or virtual human) during the function call.

The APIs for functions for the virtual human and the robot are developed in such a way that the functions are kept as similar as possible so that the same functions result in similar behaviors in the agents within the mechanical and physical limits of the hardware for each function. For example, "look at something" is a function, and if it is called for the virtual human, it will show a posture looking at something in the virtual environment. Similarly, if this function is called for the robot, it will show a similar posture of looking at something at the physical environment. It is expected that the unified platform may help the agents to be integrated and collaborate with each other to perform personal assistance to humans easily.

9.5 Home-Based Settings to Assist Disabled Persons in Daily Living by the Robot, the Virtual Human and Their Collaboration

Three home-based representative settings to assist disabled persons in their daily living for a particular task were developed as follows:

i. An intelligent humanoid robot assists a disabled person in his/her daily living activities as shown in Fig. 9.3a.
ii. An intelligent virtual human assists a disabled person in his/her daily living activities as shown in Fig. 9.3b.
iii. The robot and the virtual human collaborate with each other to assist a disabled person in his/her daily living activities as shown in Fig. 9.3c.

9.5.1 *The Intelligent Robot Assists the Human*

As Fig. 9.3a shows, 10 rectangular paper boxes (black appearance) are kept in a room, 5 boxes are put in the left side (on a table), and the remaining 5 boxes are put in the right side (on the sofas). A small object (say, a coffee mug) is hidden (by the experimenter) inside any of the 10 boxes. An intelligent robot and a human stand facing each other in the room. Let us assume that the human is physically disabled and has limited mobility. It is also assumed that the disabled human needs the coffee mug, but the current location of the mug is unknown to him/her and he/she expects that the robot will find out the mug for him/her so that he/she can get the mug quickly and easily despite his/her limited mobility. This is the way how the robot serves the human as his/her personal assistant.

The experimenter hides the object inside any of the 10 boxes and inputs the information of the position coordinates of the object/box (e.g., $O_r = [x, y, z]^T$ in

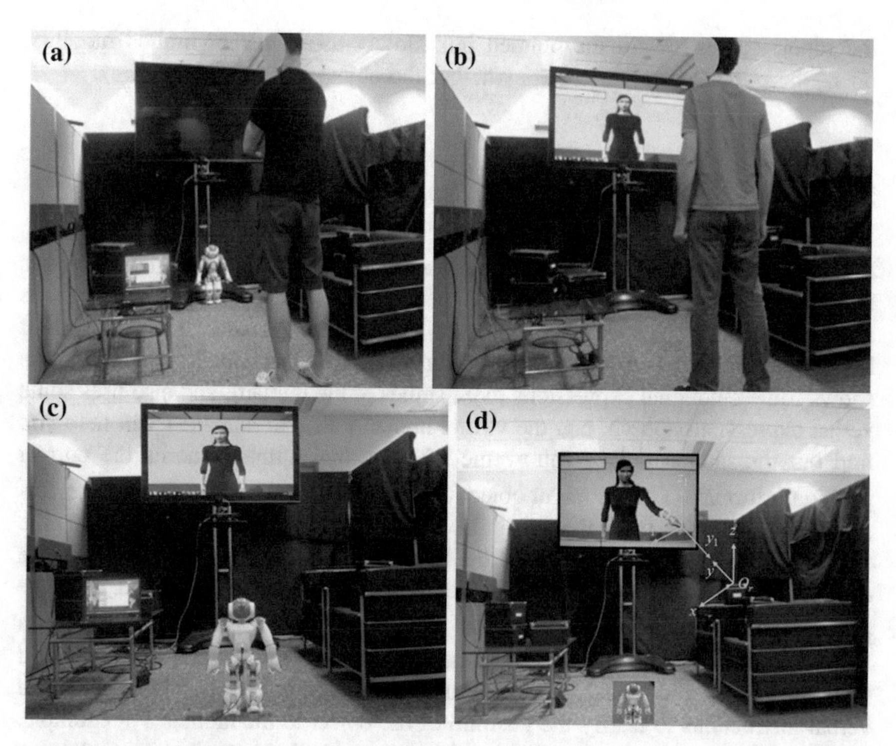

Fig. 9.3 **a** The intelligent robot assists the human, **b** the intelligent virtual human assists the human, **c** collaboration between the robot and the virtual human to assist the human, **d** an example of mapping between the pointing (fingertip) position of the virtual human in the virtual environment and the corresponding position of a targeted box in the physical environment

Fig. 9.3d) to the computer system. The robot uses the Kinect camera to recognize the head of the human. Thus, the robot understands the human's presence and his/her intent to receive service from the robot to find out the hidden/missing object (mug). The robot at first shows some gestures such as it stands straight, looks at the human standing in front of it based on head tracking, shows gaze/attention at the human, etc. Then, it shows some emotional expressions such as smiles at the human, and uses verbal expressions (speech), e.g. the robot tells, "hi human! I will help you find out the object, follow me". The robot inherits the actual position coordinate information of the target object/box in the real-world, e.g. $O_r = [x, y, z]^{\mathrm{T}}$ from the computer system. Based on this information, the robot turns its face toward the correct location of the box containing the object, moves (walks) toward the target box, stops near the target box and points at the box. Then, the robot based on pre-stored information uses additional verbal instructions to clarify the position of the hidden object to the human. For example, the robot tells, "hi human, the object is inside that box, the box containing the object is lying on a sofa closer to the screen, it is on top of another box", and so forth. In fact, the robot cannot open the box due to limitation of its skills. Hence, the human follows the instructions of the robot, moves toward the pointed box slowly using his/her limited mobility, opens the pointed box and checks whether the object exists inside the box.

9.5.2 *The Intelligent Virtual Human Assists the Human*

As Fig. 9.3b shows, the virtual human follows similar procedures as the robot uses to assist the human. The virtual human appears in the screen, shows similar gestures as the robot shows such as stands straight, looks at the human standing in front of her based on head tracking, shows gaze/attention at the human, etc. Then, she shows some emotional expressions (e.g., smiles at the human), and also uses some verbal expressions/speech, e.g. the virtual human tells, "hi human! I will help you find out the hidden object, follow me". The virtual human inherits the correct position information of the target object/box in the real-world, e.g. $O_r = [x, y, z]^T$ as shown in Fig. 9.3d from the computer system. Based on this information, the virtual human turns her face toward the correct location of the box containing the object, moves (walks) toward the target box (within the screen), and uses the pre-specified position and posture of her fingertip (e.g., $O_v = [x_1, y_1, z_1]^T$ as Fig. 9.3d shows) in the virtual environment appropriate for pointing the real-world target box. Then, the virtual human based on pre-stored information uses additional verbal instructions to clarify the position of the object to the human. For example, the virtual human tells, "hi human, the object is inside that box, the box containing the object is lying on a sofa closer to the screen, it is on top of another box", and so forth.

The virtual human cannot open the box as it cannot come outside the screen. Hence, the human based on the virtual human's instructions moves slowly toward the target box, opens the pointed box on behalf of the virtual human and checks whether the object exists inside the box. In the future, more allied technologies may be available as peripheral devices of the virtual human that may be used to open the box with the discretion of the virtual human.

9.5.3 *Collaboration Between the Robot and the Virtual Human to Assist the Human*

To find out the hidden object through the collaboration between the artificial agents (virtual human and humanoid robot), one agent (either the robot or the virtual human) acts as the master and another agent acts as the follower. Let us consider the case first when the virtual human acts as the master and the robot acts as the follower. Then, the opposite may be considered.

9.5.3.1 The Virtual Human Is the Master and the Robot Is the Follower

As Fig. 9.3c shows, during the collaboration, the virtual human appears in the screen, shows some gestures (e.g., stands straight, looks at the robot standing in front of her based on head tracking, shows gaze/attention at the robot), shows some emotional expressions (e.g., smiles at the robot) and also uses some verbal expressions (speech), e.g. the virtual human tells, "hi robot! I will help you find out the hidden object, follow me". Being the master, the virtual human inherits the position information of the target box in the real-world, e.g. $O_r = [x, y, z]^T$ as in Fig. 9.3d from the computer system. Based on this information, the virtual human turns her face toward the correct location of the box containing the object, moves (walks) toward the target box, and uses the pre-specified position and posture of her fingertip (e.g., $O_v = [x_1, y_1, z_1]^T$) in the virtual environment appropriate for pointing the real-world target box as shown in Fig. 9.3d. Then, the virtual human based on pre-stored information uses verbal instructions to clarify the position of the object to the robot, e.g. the virtual human tells, "hi robot, the object is inside that box, the box containing the object is lying on a sofa closer to the screen, it is on top of another box", and so forth.

Then, the follower agent (robot) tries to identify the correct location of the box based on the instructions it receives from the master agent (virtual human). The virtual human's fingertip position in the virtual environment is shared to the robot through the computer system, which helps the robot determine the corresponding position of the target box in the real-world space based on the preplanned mapping (Fig. 9.3d). Once the target position is determined, the robot uses some verbal expressions such as it tells, "hi virtual human! thank you for instructing the location of the object, now I may try to find it". Then, the robot shows some gestures such as turns its face toward the target position, walks to near the target, stops walking, looks at the target position (box), points at the box and tells, "I have found the box where the object may exist, thank you virtual human for your help". In fact, the robot cannot open the box due to limitation of its skills. Hence, the human slowly moves to the pointed box, opens the box and checks whether the object exists inside that box.

9.5.3.2 The Robot Is the Master and the Virtual Human Is the Follower

Similar story happens if the robot is the master and the virtual human is the follower. At the beginning, the robot uses similar gestures and verbal expressions as the virtual human uses during its role as the master agent. Being the master, the robot inherits the position information of the target box in the real-world, e.g. $O_r = [x, y, z]^T$ as in Fig. 9.3d. Then, the robot moves to near the target position and points at the target box. An Inertial Measurement Unit (IMU) attached onto the fingers of the robot hand measures the position of the fingertip, which is then passed to the virtual human through the computer system. Then, the corresponding fingertip position of the virtual human is determined based on the preplanned mapping (Fig. 9.3d), which is the target position for the virtual human. Then, the virtual human follows the instructions of the robot in the similar way as the robot (as a follower) follows the instructions of the virtual human, moves toward the target position (within the screen), points from the virtual world at the target box in the physical world as illustrated in Fig. 9.3d, and then tells, "hi robot! I have found the box, the box is that one where the object may exist". The virtual human cannot open the box as it cannot come outside the screen. Hence, the human moves to the box that was pointed jointly by the robot and the virtual human, opens the pointed box and checks whether the object exists inside that box.

Remark 1: Here, the robot and the virtual human's gestural, emotional and verbal expressions are used to mimic human-human interactions to just make the collaboration more natural.
Remark 2: The human may himself/herself find out the hidden (missing) object instead of receiving supports from the virtual human and the robot. However, let us consider the following cases where the proposed assistance may be necessary:

i. The human has no knowledge of the correct location of the hidden object.
ii. The human is disabled with limited mobility or busy and thus wants to get the object searched out by the agents or through their collaboration (assisted living). In this case, the human still needs to move to pick the object, but limited movement may be sufficient to get the object as the human knows the correct location of the object through the assistance of the agents. The human may not need to move at all if the artificial agents are enough capable to open the box and carry the object to the immobile human, which is not feasible now but may be feasible in the near future.
iii. The human may be partly cognitively disabled and hence cannot use his/her intelligence to process visual and cognitive information to determine the search path, and the human may have problem in memory.
iv. The human actually does not need to command the robot or the virtual human to start searching the object. Instead, the robot or the virtual human can understand the human's intent through head recognition (recognition of presence) based on ambient information (visual information through the Kinect). This is suitable for cognitively disabled humans who even cannot express their needs or cannot command someone to receive help.

Remark 3: One agent (either the virtual human or the robot) may find out the object alone without collaborating with another one. However, let us consider the following cases where such collaboration may be necessary:

i. One agent has no knowledge about the correct location of the hidden object.
ii. One agent who has knowledge about the object is not physically present in the site, and thus appears through telepresence to help another local agent to find out the object.

iii. One agent is less intelligent, but more physically skillful to find out the object, and the vice versa.
iv. As both agents are artificial with limited skills, intelligence and capabilities, collaboration between them may benefit each other with complementary attributes, intelligence and skills, which may ease the task and enhance the reliability of the assistance provided to the human.

9.5.4 Strategy of Determining the Master and the Follower Agent

Whether the virtual human or the robot should act as the master agent depends on the bilateral trust between them. The collaboration scheme including the switching of master (leader)-follower role based on the bilateral trust is given in Fig. 9.4, where $T_{VH2HR}(prior\ trial)$ is the virtual human's trust in the robot in immediate prior trial (a run), and $T_{HR2VH}(prior\ trial)$ is the robot's trust in the virtual human in that immediate prior trial. Usually, the master takes the initiative to find out the hidden object, and hence the initiatives also switch between the agents depending on their bilateral trust. Thus, the collaboration may be considered as a mixed-initiative collaboration [30] triggered by bilateral trust between the artificial agents.

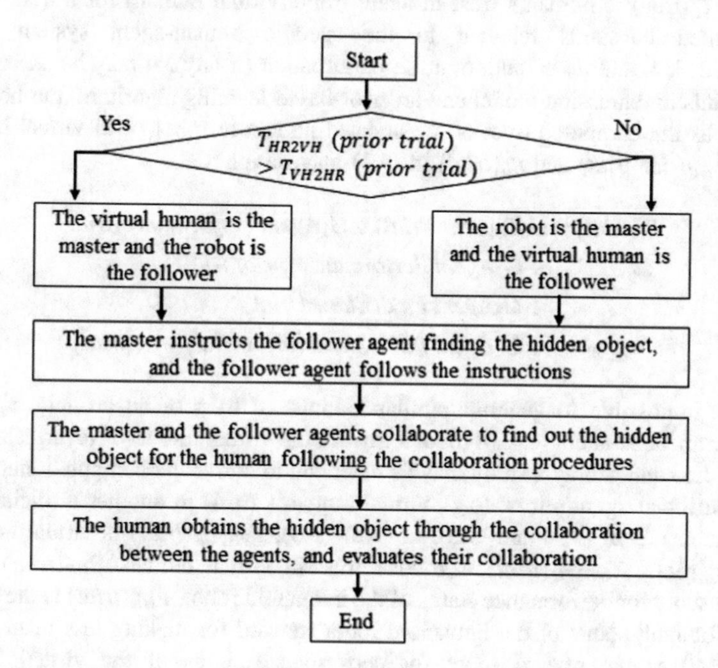

Fig. 9.4 The collaboration scheme and the switching of master (leader)-follower role based on the bilateral trust of the robot and the virtual human

9.6 Modeling and Measurement of Human Trust in Robot and Virtual Human and Bilateral Trust Between Robot and Virtual Human

Modeling and measurement of trust is an enabling step to implement the proposed collaboration between the robot and the virtual human explained in Sect. 9.5.3. The switch of their roles in taking initiatives as proposed in Sect. 9.5.4 is also based on the bilateral trust between them. In order to support the proposed collaboration, the trust modeling and measurement methods are presented below.

9.6.1 Trust Modeling

Trust is actually a perceptual issue and the human has actual feeling of trust in an artificial agent or in another human. A computational model of human's trust in the robot or in the virtual human is proposed here. Though trust of one agent in another agent may depend on many factors, in Lee and Moray's study [45], a time-series model based only on performance and faults of automation (artificial agent) was used to compute human's (biological agent) trust in automation. A general computational model of human trust in robot/virtual human may be expressed in (9.1), where, T_a(trial) is human's trust in agent (robot/virtual human) for a trial, G_i are real-valued constants relevant to the specific human-agent system where $i = 0, 1, 2, 3, 4$, and n_p is random noise perturbation (if any). It may be an ordinary deterministic regression model and an error-based learning algorithm, but here it is treated as the computed trust of the assisted human in robot or in virtual human. Here, $T_a(\mathit{prior\ trial})$ and $n_p(\mathit{trial})$ in (9.1) may be ignored.

$$\begin{aligned} T_a(\mathit{trial}) = {} & G_0 T_a(\mathit{prior\ trial}) + G_1 \mathit{Agent\ Performance}(\mathit{trial}) \\ & + G_2 \mathit{Agent\ Performance}(\mathit{prior\ trial}) \\ & + G_3 \mathit{Agent\ Fault\ Status}(\mathit{trial}) \\ & + G_4 \mathit{Agent\ Fault\ Status}(\mathit{prior\ trial}) + n_p(\mathit{trial}) \end{aligned} \quad (9.1)$$

It is impossible to generate similar feelings of trust of an artificial agent in another artificial agent (e.g., a robot's trust in the virtual human). Nevertheless, the idea in Lee and Moray's study may be extended to derive the computational model of an artificial agent's trust (e.g., virtual human's trust) in another artificial agent (e.g., robot) as in (9.2) and (9.3). In (9.2)–(9.3), $T_{VH2HR}(\mathit{trial})$ is virtual human's trust in robot, $T_{HR2VH}(\mathit{trial})$ is robot's trust in virtual human, $P_{HR}(\mathit{trial})$ is the reward score for performance status of the humanoid robot, $F_{HR}(\mathit{trial})$ is the reward score for fault status of the humanoid robot (reward for making less or no fault), $P_{VH}(\mathit{trial})$ is the reward score for performance status of the virtual human, $F_{VH}(\mathit{trial})$ is the reward score for fault status of the virtual human, and $n_{p1}(\mathit{trial})$

and $n_{p2}(trial)$ are the noise and perturbation in a trial, and A_i and B_i are the real-valued constants $(i = 1, 2, 3, 4)$ that depend on the specific tasks and agents. Here, $n_{p1}(trial)$ and $n_{p2}(trial)$ may be ignored.

$$T_{VH2HR}(trial) = A_1 P_{HR}(trial) + A_2 P_{HR}(prior\ trial) + A_3 F_{HR}(trial) + A_4 F_{HR}(prior\ trial) + n_{p1}(trial) \tag{9.2}$$

$$T_{HR2VH}(trial) = B_1 P_{VH}(trial) + B_2 P_{VH}(prior\ trial) + B_3 F_{VH}(trial) + B_4 F_{VH}(prior\ trial) + n_{p2}(trial) \tag{9.3}$$

9.6.2 Trust Measurement

In order to measure T_a for a trial, the assisted human uses a five-point Likert scale to rate his/her evaluation of the agent performance and the agent's fault status for a trial. The Likert scale is shown in Fig. 9.5. The agent performance is expressed in a few criteria such as the agent's speed (efficiency) in assisting the human finding the object. The agent fault status is expressed in term of the agent's accuracy in pointing the box containing the object. For example, the human responds the following two questions separately using the Likert scale:

i. How was the performance of the agent in assisting you in finding the object?
ii. How was the fault avoiding ability or the accuracy of the agent in assisting you in finding the object?

For real-time measurements of $T_{VH2HR}(trial)$ and $T_{HR2VH}(trial)$, real-time measurements of performance and fault status of the robot and the virtual human are necessary. Movement speed of the robot from its initial position to the target position is considered as a criterion of robot performance, and the deviation of the actual position pointed by the robot fingertip from the target position is considered as a criterion of robot fault. Ideally, the robot speed and the target position of the fingertip are fixed (set by computer program), and thus there should have no deviation. However, external disturbances (e.g., friction between floor surface and robot feet during walking, resistance of air, etc.) and change in robot stiffness due to heat generated by the actuating motors may be the reasons behind the potential

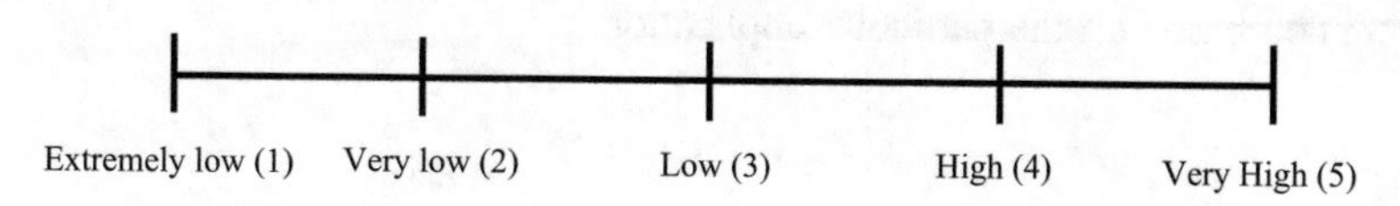

Fig. 9.5 Likert scale to measure human's trust in the agent for an assistance trial

deviations in robot path trajectory and speed, which may cause the deviations between the actual and the target pointing positions. An IMU attached onto robot's hand fingers can be used to measure its actual fingertip position when it points at the target box. An IMU set over robot's foot may measure its walking speed. Let V_{rm} is the measured (actual) absolute walking speed of the robot, V_{rs} is the set (desired) absolute speed for the robot, d_{rs} is the absolute shortest distance between the target position (position of the target box) and the actual position of the robot fingertip when it points at the box. The target position for a box is considered as the center of the top view of the box if it is not stacked with another box. The center of the front side (facing the robot) of the box is the target position for the box if the box is stacked with other boxes. δ is a small magnitude of threshold determined based on particular application. $P_{HR}(trial)$ and $F_{HR}(trial)$ may be objectively msured at real-time using (9.4)–(9.5).

$$P_{HR}(trial) = \begin{cases} 1.00, & if\ V_{rm} \approx V_{rs} \\ 0.75, & if\ V_{rs} > V_{rm} \geq 0.75\,V_{rs} \\ 0.50, & if\ 0.75\,V_{rs} > V_{rm} \geq 0.5\,V_{rs} \\ 0.25, & if\ 0.5\,V_{rs} > V_{rm} \geq 0.25\,V_{rs} \\ 0.00, & if\ V_{rm} < 0.25\,V_{rs} \end{cases} \tag{9.4}$$

$$F_{HR}(trial) = \begin{cases} 1, & if\ d_{rs} \approx 0 \\ 0.5, & if\ d_{rs} \leq \delta \\ 0.0, & otherwise \end{cases} \tag{9.5}$$

Similarly, the movement speed and deviation of the actually pointed position from the target pointing position of the virtual human are considered as its performance and fault criterion respectively. Same as the robot, the speed and the target pointing position of the virtual human are ideally fixed, but the communication delay between the client and the control server, disturbances, software errors and system instabilities may reduce the speed of the virtual human. All these including other uncertainties such as calibration (mapping) errors, unnoticed displacement of the screen or the box, etc. may cause deviation in virtual human's pointed position from target position. If V_{vhm} is the actual absolute speed of the virtual human (obtained through the computer system for the virtual human), V_{vhs} is the absolute speed set for the virtual human, d_{vhs} is the absolute shortest distance between the target (set) position for the fingertip in the virtual environment appropriate for pointing the corresponding real-world box (e.g., O_v in Fig. 9.3d) and the actual position of the virtual human's fingertip in the virtual environment (obtained through computer system), then $P_{VH}(trial)$ and $F_{VH}(trial)$ may be measured at real-time following (9.6)-(9.7), where σ is a small magnitude of threshold determined based on the particular application.

$$P_{VH}(trial) = \begin{cases} 1.00, & if\ V_{vhm} \approx V_{vhs} \\ 0.75, & if\ V_{vhs} > V_{vhm} \geq 0.75\ V_{vhs} \\ 0.50, & if\ 0.75\ V_{vhs} > V_{vhm} \geq 0.5\ V_{vhs} \\ 0.25, & if\ 0.5\ V_{vhs} > V_{vhm} \geq 0.25\ V_{vhs} \\ 0.00, & if\ V_{vhm} < 0.25\ V_{vhs} \end{cases} \tag{9.6}$$

$$F_{VH}(trial) = \begin{cases} 1, & if\ d_{vhs} \approx 0 \\ 0.5, & if\ d_{vhs} \leq \sigma \\ 0.0, & otherwise \end{cases} \tag{9.7}$$

Note 1: δ and σ need to be adjusted as the chance of the robot being affected by external disturbances is higher than that of the virtual human. One approach is to consider $\delta \gg \sigma$, which may help compare the faults between the agents on similar extent.

Note 2: P_{VH} and P_{HR} are between 0 (least performance) and 1 (maximum performance), and F_{VH} and F_{HR} are between 0 (there is some fault) and 1 (there is no fault). In fact, P_{VH} and P_{HR} mean the reward for showing good performance, and F_{VH} and F_{HR} mean the reward for making no or less fault. These values may be normalized between 1 and 5 to make the computed trust values (T_{VH2HR} and T_{HR2VH}) comparable with the human's assessed trust values obtained using the Likert scale.

Note 3: The trust models may be verified using model verification tools [29], but the models are validated here using experimental evaluation results, as follows.

9.7 Evaluation Scheme to Evaluate the Assistance of the Robot, Virtual Human and Their Collaboration to the Disabled Human

The quality of assistance of the robot, virtual human and their collaboration to disabled humans is expressed in three categories as follows: (i) interaction quality between the human and the robot, the virtual human and their collaboration, (ii) overall task performance, and (iii) human's overall likeability of the assistance.

Interaction quality between the human and the robot, the virtual human and their collaboration is evaluated using two sub-categories of criteria: (a) attributes of the assistant agent, (b) effectiveness of interactions between the human and the agents. The attributes of the agent are evaluated based on the following terms: (i) level of anthropomorphism of the agent, (ii) level of agent embodiment, (iii) quality of verbal and facial expressions, gestures, actions and instructions of the agent, and (iv) level of stability of the agent. The effectiveness of interactions between the human and the agent is evaluated based on the following terms: (i) cooperation level, (ii) level of engagement between the human and the agent, (iii) naturalness (similarity with human-human interactions) in the interactions, (iv) prospect/potential of long term companionship between the human and the agent, and (v) situation awareness of the assisted human. The attributes and the effectiveness of interactions are subjectively evaluated by the assisted human using the Likert scale shown in Fig. 9.5.

The task performance is evaluated objectively following two criteria: (i) efficiency, and (ii) success rate in finding the hidden object. The efficiency and success rate are measured following (9.8) and (9.9) respectively. In (9.8), T_t is the targeted time to complete the search task for a trial, and T_A is the actual time for the task. In (9.9), n_{tf} is the total number of collaboration trails where the human could not obtain the hidden object based on the instructions received from the agents or from the collaboration of the two agents, and n_t is the total number of trails.

$$\mathit{Efficiency}(\eta) = \left(\frac{T_t}{T_A}\right) \times 100\% \tag{9.8}$$

$$\mathit{Success\ rate}(\lambda) = \left(1 - \frac{n_{tf}}{n_t}\right) \times 100\% \tag{9.9}$$

Human's likeability of the assistance received from the individual agents and their collaboration is expressed through the levels of human's (i) satisfaction in the assistance, (ii) own trust in the agent/collaboration, and (iii) dependability in the service. These criteria are assessed by the human using the Likert scale in Fig. 9.5.

9.8 Experimental Evaluation of the Quality of the Assistance of the Robot, the Virtual Human and Their Collaboration to the Disabled Human

9.8.1 Recruitment of Subjects

Thirty (30) human subjects (engineering students, males 26, females 4, mean age 25.67 years with variance 3.58 years) were recruited to participate in the experiment. All the subjects reported to be physically and mentally healthy. The study was approved by the concerned ethical committee.

9.8.2 Experimental Objectives

The objectives of the experiment were to evaluate the quality of the assistance provided to disabled humans in finding the object by the robot, the virtual human and their collaboration following the evaluation scheme introduced in Sect. 9.7.

9.8.3 Hypotheses

It was hypothesized that-

i. **Hypothesis I**: The virtual human and the robot operated through the common platform based on similar functions (APIs) would show similar satisfactory behaviors and performance in assisting the humans.
ii. **Hypothesis II**: More human-like attributes of the artificial assistant agents would be perceived better by the assisted humans and would result in better interaction quality and assisted human's likeability.
iii. **Hypothesis III**: Collaboration between assistant agents would result in better perceived attributes, performance and assisted human's likeability than that resulted in for individual assistant agents (robot or virtual human).

9.8.4 Experimental Procedures

At first, a few practice trials for assisting the humans by the robot, the virtual human and their collaboration as shown in Figs. 9.3a-c respectively were implemented. The information on agent performance and faults for the assistant robot and the virtual human and their collaboration obtained during the practice sessions were used to compute the constants G_i, A_i and B_i of the trust models in (9.1)–(9.3) respectively following the Autoregressive Moving Average Model (ARMAV) [29], as given in Table 9.1. The necessary thresholds to compute the trust were also decided (Table 9.1).

Then, the experiment sessions started. During the experiment sessions, the robot assisted a human subject finding out the missing household object as depicted in Fig. 9.3a, which is named as "robot assists" assistance protocol. Then, the human subject evaluated the interaction quality and his/her likeability of the assistance following the evaluation scheme (Sect. 9.7). The human subject also evaluated the agent performance and agent fault status subjectively. The experimenter evaluated the task performance following the evaluation scheme. The evaluation data were recorded properly. Then, the subject was replaced by another subject, and in this way the 30 recruited subjects completed 30 trials separately.

Table 9.1 Values of the constants for trust computation

Constant	Value	Constant	Value	Constant	Value
G_1	0.4902	A_1	0.513	B_1	0.487
G_2	0.050	A_2	0.052	B_2	0.061
G_3	0.4201	A_3	0.396	B_3	0.403
G_4	0.0397	A_4	0.039	B_4	0.049
		δ	0.005	σ	0.001

In the similar way, the 30 subjects participated separately in the cases when the virtual human as in Fig. 9.3b assisted the human in finding out the missing object (named as "virtual human assists" assistance protocol), and their collaboration as in Fig. 9.3c assisted the humans in finding out the missing object (named as "collaboration assists" assistance protocol). Similar experimental procedures as utilized for the case when the robot assisted the human in Fig. 9.3a were utilized for the cases when the virtual human and the collaboration assisted the human. The same 30 subjects participated in three experiment protocols. However, the protocols were randomized and the object was randomly hidden in different boxes that removed or reduced the learning effects of the subjects.

9.8.5 Experimental Results

Figure 9.6 shows that the attributes of the robot as a personal assistant were perceived better by the assisted humans than those of the virtual human as a personal assistant to the human. It is assumed that human-like 3-dimensional physical existence of the robot generated better perceptions of its overall attributes in the human than in its virtual counterpart, which justifies hypothesis II. The results also show that the attributes perceived by the assisted human were better for the collaboration of two agents than for the individual agents. It might happen because the human perceived the attributes of both physical and virtual agents during the collaborative assistance together, which might generate a better perception of the agent attributes than the attributes the human perceived when individual agents assisted the human. The results thus justify hypothesis III. Analysis of Variances (ANOVAs) showed that variations in attribute scores between subjects for each attribute criterion for the three assistance protocols were statistically nonsignificant (e.g., for level of agent anthropomorphism, $F_{29,58} = 1.44, p > 0.05$.), which indicate the generality of the results. However, variations in attribute scores among the three assistance protocols were statistically significant ($F_{2,58} = 23.19, p < 0.05$), which statistically prove that the subjects perceived the attributes of the artificial agents for different assistance protocols differently. However, there were slight different results for the stability. The stability of the robot was lower than that of the virtual human probably due to the reasons that the robot was affected by the disturbances such as floor roughness, motor temperature, obstacles, air resistance, etc., but the virtual human was not so affected by those disturbances. The robot stability improved when the assistance was provided by the collaboration, but it was still lower than that when the assistance was provided by the virtual human. It might happen because the comparatively less stable robot was involved in the collaboration with the comparatively more stable virtual human, which might reduce the resultant stability of their collaboration. Nevertheless, the attributes of both artificial agents and their collaboration were perceived as satisfactory according to the scale in Fig. 9.5 by the assisted humans. The results thus justify the effectiveness of generating human-like satisfactory attributes in artificial agents of heterogeneous

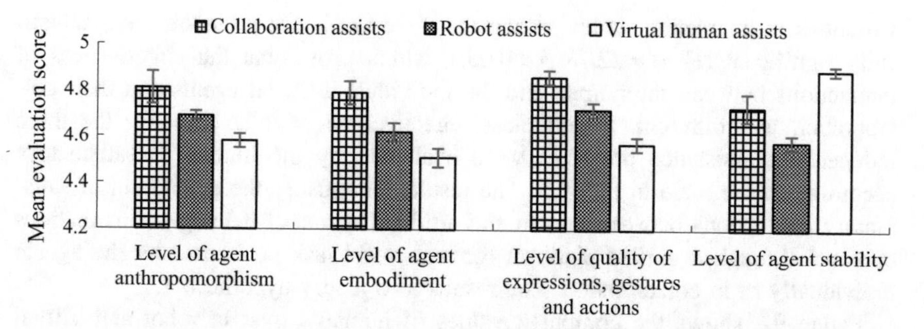

Fig. 9.6 Evaluation results of attributes of the assistant agents for different assistance protocols

realities operated through a unified platform for the real-world task performed individually or in collaboration for assistance to humans. The results thus justify hypothesis I.

Figure 9.7 shows that the effectiveness of interaction of the robot as a personal assistant was rated higher by the assisted humans than that of the virtual human as a personal assistant to the humans. It is assumed that human-like 3-dimensional physical existence of the robot helped produce better interactions over its virtual counterpart, which justifies hypothesis II. The results also show that the effectiveness of interaction perceived by the assisted humans were better for the collaboration of the agents than that for the individual agents. It might happen because the human received the assistance of both the physical and virtual agents during the collaborative assistance together that might generate a better interaction environment than the interactions the human experienced when individual artificial agents assisted the human. The results thus justify hypothesis III. The interaction results were also in line with the perceived attributes. ANOVAs showed that variations in interaction scores between subjects for each interaction criterion for the three assistance protocols were statistically nonsignificant (e.g., for level of cooperation, $F_{29,58} = 1.39, p > 0.05$), which indicate the generality of the results. However,

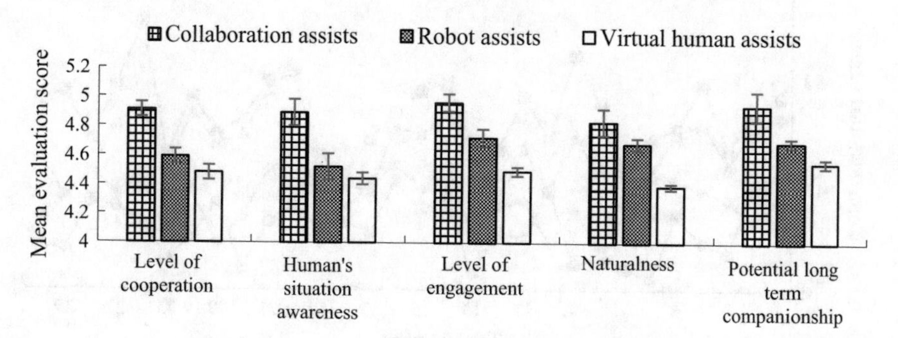

Fig. 9.7 Evaluation results for effectiveness of interaction for different assistance protocols

variations in interaction scores among the three assistance protocols were statistically significant ($F_{2,58} = 22.78, p < 0.05$.), which prove that the effectiveness of interactions between the human and the individual artificial agents and their collaboration was different. Nevertheless, effectiveness of interaction for the three independent assistance protocols were evaluated by the human as satisfactory according to the scale in Fig. 9.5. The results thus justify the generation of satisfactory interactions between human and artificial agents of heterogeneous realities operated through a unified platform for real-world task performed by the agents individually or in collaboration. The results also justify hypothesis I.

Figure 9.8 shows the computed values of human's trust in robot and virtual human, robot's trust in virtual human and virtual human's trust in robot for different trials. The results show that the trust of the human on the artificial agents and the trust between the artificial agents were high (at least 4 out of 5 or 80% and above), which indicate that the human was willing to receive assistance from the artificial agents, and the assistance of one artificial agent toward another artificial agent was also rational and practical for the selected collaboration task. The high trust on the artificial agents also proved the ability of the artificial agents to produce high performance and avoid faults during assisting the human individually or collaboratively [28]. The results thus justify the effectiveness of generating similar and satisfactory skills and capabilities in the artificial agents of heterogeneous realities for a real-world common task through a common platform, and also justify hypothesis I. The results show that human's trust in the virtual human was higher than that in the robot even though the attributes of the robot perceived by the human (Fig. 9.6) and the effectiveness of interaction (Fig. 9.7) were better for the robot than that for the virtual human. It might happen due to comparatively slower motion and less accuracy in motion and action of the robot caused by the disturbances of the physical world than comparatively faster and more accurate motion and action

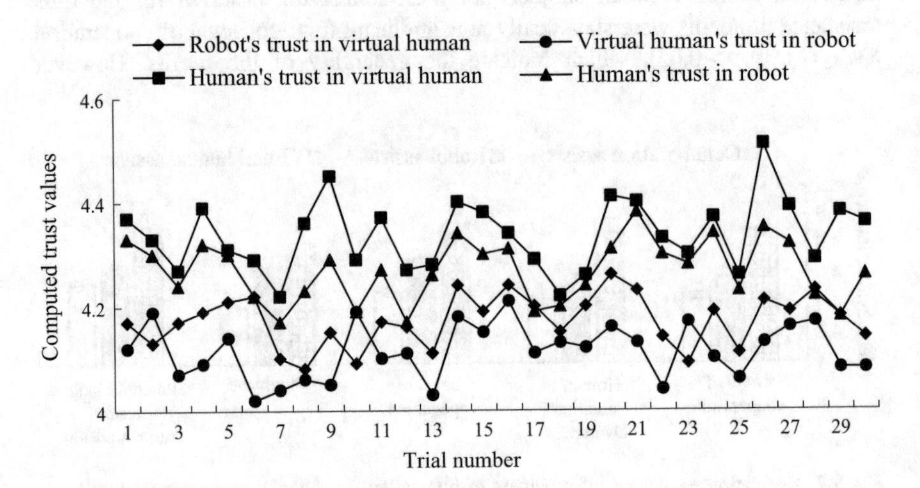

Fig. 9.8 Computed trust values for different protocols

of the virtual human in the virtual environment that was less affected by external disturbances. The results also show that the robot's trust in the virtual human was higher than the virtual human's trust in the robot for 27 trials out of 30 trials (i.e., in 90% cases). It indicates that the virtual human served as the master agent and the robot served as the follower agent in 90% trials under the protocol when the artificial agents assisted the human collaboratively. Again, in the remaining 10% trials, the role of the agents as master and follower as well as their turns in taking initiatives were switched based on the bilateral trust values according to the collaboration scheme (Fig. 9.4). The results thus also prove the effectiveness of the mixed-initiatives in the collaboration of two artificial agents of heterogeneous realities for personal assistance to humans [30]. The results also show that the assisted human's trust was the highest when the virtual human assisted the human. The trust then decreased when the robot assisted the human, and the trust further decreased when their collaboration assisted the human. It was also compatible with the results that the robot's trust in the virtual human was higher than the virtual human's trust in the robot.

Figure 9.9 shows that the efficiency was the highest when the virtual human assisted the human. The efficiency dropped when the robot assisted the human. The efficiency further dropped when their collaboration assisted the human. The results were in line with the computed trust levels in Fig. 9.8. It might happen due to comparatively slower motion and action of the robot caused by the disturbances of the physical world (floor roughness, motor temperature, obstacles, air resistance, etc.) than comparatively faster motion and action of the virtual human in the virtual environment, which was less affected by external disturbances. The efficiency slightly dropped and became less uniform for the protocol when the assistance was provided by the collaboration because the comparatively less efficient robot was involved in the collaboration with comparatively more efficient virtual human,

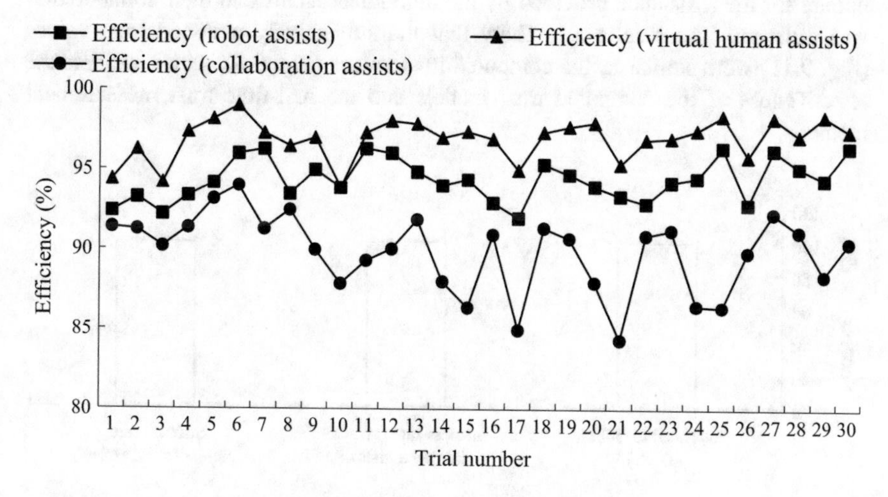

Fig. 9.9 Task efficiencies in the trials for different assistance protocols

which reduced the resultant efficiency of the collaboration. Nevertheless, levels of efficiency for the three independent assistance protocols were satisfactory (above 80%).

Figure 9.10 shows that despite having slight differences in perceived attributes, effectiveness in interactions, trust levels and efficiencies, the assistance was 100% successful for all assistance protocols. The results thus justify the effectiveness of generating efficient and successful interactions between human and artificial agents of heterogeneous realities for real-world task performed by the agents individually or in collaboration operated through the common platform.

Figure 9.11 shows the likeability of the humans for the assistance they received from the artificial agents and their collaboration for the selected task. The results show that the assisted humans were satisfied with the assistance and thus their own trust toward the agents and their collaboration was also high. High level satisfaction and trust of the assisted humans indicate their interest and willingness to receive the assistance and show their dependency on the assistance provided by the agents in three independent protocols. The results show that the likeability was higher when the robot assisted the human than when the virtual human assisted. The likeability dropped when the virtual human assisted the human even though the virtual human had better efficiency and trustworthiness over the robot. Better attributes and interactions with the human due to having 3-dimensional physical existence of the robot might be the reason of more likeability of the robot over its virtual counterpart. Again, humans liked the assistance provided by the collaboration of the assistant agents more than that provided by the individual agents. ANOVAs showed that variations in likeability scores between subjects for each likeability criterion for the three assistance protocols were statistically nonsignificant (e.g., for satisfaction, $F_{29,58} = 1.33, p > 0.05$), which indicate the generality of the results. However, variations in likeability scores among the three assistance protocols were statistically significant ($F_{2,58} = 21.54, p < 0.05$), which proved that the likeability of the human for the assistance provided by the individual agents and their collaboration was different. The results also show that the subjectively assessed trust values (Fig. 9.11) were similar as the computed trust values (Fig. 9.8), which validate the effectiveness of the computed trust models and the real-time trust measurement methods.

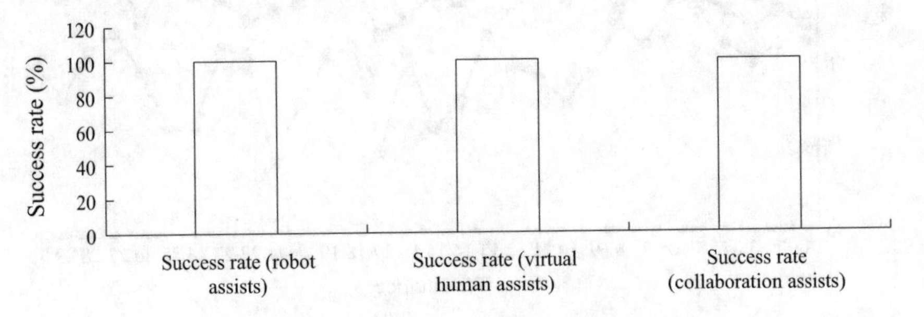

Fig. 9.10 Success rates for different assistance protocols

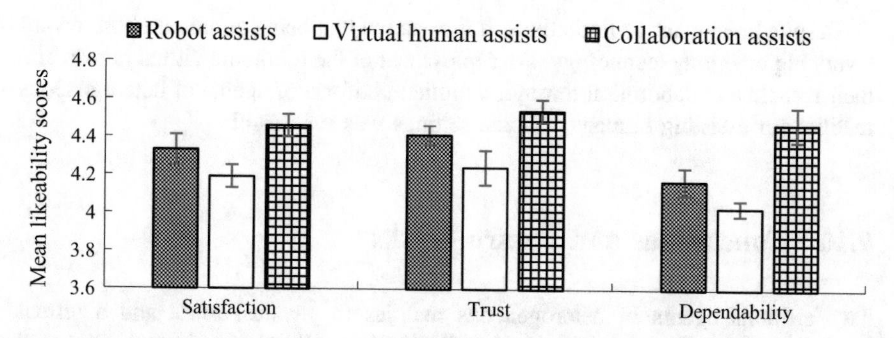

Fig. 9.11 Human's likeability of assistance received for different assistance protocols

The above results generally show that, being a physical agent, the robot had better attributes and interaction quality than the virtual human, but the robot's lower stability and vulnerability to disturbances compared to its virtual counterpart might reduce its performance, accuracy, efficiency, and trustworthiness, which reduced its role in taking initiatives in the mixed-initiative collaboration. Despite having slight differences in attributes, interaction quality, stability, performance, accuracy and efficiency among the robot, the virtual human and their collaboration, the fact is that the real-world assistance to the humans by the artificial agents and their collaboration through a common platform was successful. All the three adopted hypotheses were justified and were proved to be true. The results thus pave the way to applying the artificial agents of heterogeneous realities for various purposes such as assisted living, social companionships for old and lonely people, personal assistance, etc.

9.9 Limitations of the Methods and the Results

- The activities of the virtual human and the robot were pre-planned and thus limited. For example, the agents performed their actions to assist the human and each other only in pre-planned ways for objects hidden only in a few locations (10 locations). It is assumed that the intelligence and capabilities of the agents may be enhanced so that they are more robust to assist or collaborate in many ways for objects hidden in any arbitrary location within the specified space.
- The real-time measurement of bilateral trust between the agents seemed to be less robust due to limitations in sensor arrangement. More ambient and other sensing methods may make the real-time measurement of the bilateral trust more robust, repeatable and reproducible.

Despite having above limitations, it is assumed that being an initial effort toward a very big emerging technology, the deployment of the robot, the virtual human and their real-time collaboration through a unified platform of agents of heterogeneous realities for assisting humans at home settings was successful.

9.10 Conclusions and Future Works

Two artificial agents of heterogeneous realities (a physical robot and a virtual human) were developed with a set of similar skills, intelligence and capabilities, and were integrated through a common platform to perform individually and collaboratively a real-world common task (searching for a missing object) in a homely environment as an assistance to disabled humans. Models for human's trust in the agents and bilateral trust models between the agents were derived and a trust-based collaboration scheme between the agents was developed. The levels of assistance of the individual robot, the virtual human and their collaboration were evaluated experimentally using a comprehensive evaluation scheme. The results show that the collaboration between the agents partly outperformed the individual agents in assisting the humans, and the physical agent (robot) partly outperformed the virtual agents in assisting humans in real-world home-settings. However, all three protocols (robot as an assistant, virtual human as an assistant, and their collaboration as an assistant) were satisfactory to assist the humans in the real-world settings. The major novelties and contributions of the proposed emerging technologies are: (i) intelligence, autonomy and capabilities of the virtual human and the robot were enhanced, (ii) the virtual human was empowered to perform beyond the virtual environment, (iii) a common platform was demonstrated as an emerging technology to operate agents of heterogeneous realities, (iv) human's trust in agents and bilateral trust models between artificial agents were derived and real-time measurement of bilateral trust was proposed to establish trust in artificial agents of heterogeneous realities, and (v) a comprehensive scheme was proposed to evaluate the interactions between the agents of heterogeneous realities for their assistance to disabled humans. The evaluation results justified the effectiveness of the proposed approaches of the emerging technologies. The results may be useful to develop advanced emerging technologies such as adaptive social ecologies and cyber-physical-social systems (CPSSs) using smart agents of heterogeneous realities with ambient intelligence for personal assistance in assisted living, social companionships, smart homes, etc.

In the future, agent intelligence will be enhanced to make the collaboration feasible for objects placed in any location in the space. Novel controls will be used to control the collaboration within desired specifications. The measurements of bilateral trust in real-time and the trust-based collaboration will be made more reliable, repeatable and reproducible using more appropriate sensing technologies. Personal assistance to humans will be provided in the context of more practical daily activities.

Acknowledgements The author acknowledges all the supports especially the supports for the development of the experimental facilities, artificial characters and the common platform as well as the technical supports for implementation of the robot-virtual human collaboration that he received from his ex-colleagues and research staffs of Nanyang Technological University for the research and development presented herein.

References

1. Chen, K.J., Barthes, J.P.: Enhancing intelligence of personal assistant agent using memory mechanism. In: Proceedings of 2007 11th International Conference on Computer Supported Cooperative Work in Design, Melbourne, Vic., pp. 361–365 (2007)
2. Santos, J., Rodrigues, J.J.P.C., Casal, J., Saleem, K., Denisov, V.: Intelligent personal assistants based on internet of things approaches. IEEE Syst. J. **PP**(99), 1–10 (2016)
3. Homayounvala, E., Aghvami, A.H., Groves, I.S.: On migration policies for personal assistant agents embedded in future intelligent mobile terminals. In: Proceedings of 2005 6th IEEE International Conference on 3G and Beyond, Washington, DC, pp. 1–5 (2005)
4. Koubaa, A., Sriti, M., Javed, Y., Alajlan, M., Qureshi, B., Ellouze, F., Mahmoud, A.: Turtlebot at office: a service-oriented software architecture for personal assistant robots using ROS. In: Proceedings of 2016 International Conference on Autonomous Robot Systems and Competitions (ICARSC), Bragança, Portugal, pp. 270–276 (2016)
5. Qiu, R., Ji, Z., Noyvirt, A., Soroka, A., Setchi, R., Pham, D., Xu, S., Shivarov, N., Pigini, L., Arbeiter, G., Weisshardt, F., Graf, B., Mast, M., Blasi, L., Facal, D., Rooker, M., Lopez, R., Li, D., Liu, B., Kronreif, G., Smrz, P.: Towards robust personal assistant robots: Experience gained in the SRS project. In: Proceedings of 2012 IEEE/RSJ International Conference on Intelligent Robots and Systems, Vilamoura, pp. 1651–1657 (2012)
6. Mishra, A., Makula, P., Kumar, A., Karan, K., Mittal, V.K.: A voice-controlled personal assistant robot. In: Proceedings of 2015 International Conference on Industrial Instrumentation and Control (ICIC), Pune, pp. 523–528 (2015)
7. Luria, M., Hoffman, G., Megidish, B., Zuckerman, O., Park, S.: Designing Vyo, a robotic smart home assistant: Bridging the gap between device and social agent. In: Proceedings of 25th IEEE Int. Symposium on Robot and Human Interactive Communication (RO-MAN), New York, NY, pp. 1019–1025 (2016)
8. Webster, M., Dixon, C., Fisher, M., Salem, M., Saunders, J., Koay, K., Dautenhahn, K., Saez-Pons, J.: Toward reliable autonomous robotic assistants through formal verification: a case study. IEEE Trans. Hum. Mach. Syst. **46**(2), 186–196 (2016)
9. Amato, G., Bacciu, D., Broxvall, M., Chessa, S., Coleman, S., Di Rocco, M., Dragone, M., Gallicchio, C., Gennaro, C., Lozano, H., McGinnity, T., Micheli, A., Ray, A., Renteria, A., Saffiotti, A., Swords, D., Vairo, C., Vance, P.: Robotic ubiquitous cognitive ecology for smart homes. J. Intell. Robot. Syst. **80**(1), 57–81 (2015)
10. Rahman, S.: Generating human-like social motion in a human-looking humanoid robot: the biomimetic approach. In: Proceedings of IEEE International Conference on Robotics and Biomimetics (ROBIO 2013), pp. 1377–1383, 12–14 Dec. 2013
11. Xiong, G., Zhu, F., Liu, X.. Dong, X., Huang, W., Chen, S., Zhao, K.: Cyber-physical-social system in intelligent transportation. IEEE/CAA J. Autom. Sin. **2**(3), 320–333, July 10, 2015
12. Martinez-Hernandez, U., Damianou, A., Camilleri, D., Boorman, L.W., Lawrence, N., Prescott, T.J.: An integrated probabilistic framework for robot perception, learning and memory. In: Proceedings of IEEE International Conference on Robotics and Biomimetics (ROBIO), pp. 1796–1801 (2016)
13. Hays, M., Campbell, J., Trimmer, M., Poore, J., Webb, A., Stark, C., King, T.: Can role-play with virtual humans teach interpersonal skills? In: Proceedings of Interservice/Industry Training, Simulation and Education Conference (I/ITSEC), p. 12 (2012). Paper No. 12318

14. SikLanyi, C., Geiszt, Z., Karolyi, P., Magyar1, A.: Virtual reality in special needs early education. Int. J. Virtual Real. **5**(4), 55–68 (2006)
15. Campbell, J., Hays, M., Core, M., Birch, M., Bosack, M., Clark, R.: Interpersonal and leadership skills: using virtual humans to teach new officers. In: Proceedings of Interservice/Industry Training, Simulation, and Education Conference (2011). Paper No. 11358
16. Saleh, N.: The value of virtual patients in medical education. Ann. Behav. Sci. Med. Educ. **16** (2), 29–31 (2010)
17. Lawford, P., Narracott, A., McCormack, K., Bisbal, J., Martin, C., Brook, B., Zachariou, M., Kohl, P., Fletcher, K., Diaz-Zucczrini, V.: Virtual physiological human: training challenges. Phil. Trans. R. Soc. A **368**(1921), 2841–2851 (2010)
18. Rahman, S.: Evaluating and benchmarking the interactions between a humanoid robot and a virtual human for a real-world social task. In: Proceedings of the 6th International Conference on Advances in Information Technology (IAIT 2013), Dec 12–13 (2013). Bangkok, Thailand. In: Communications in Computer and Information Science, Springer, vol. 409, pp. 184–197 (2013)
19. Rahman, S.: People-centric adaptive social ecology between humanoid robot and virtual human for social cooperation. In: Proceedings of the 2nd International Workshop on Adaptive Robotic Ecologies (ARE 2013) at the 4th International Joint Conference on Ambient Intelligence (AmI 2013), Dec 3–5, 2013, Dublin, Ireland. In: Communications in Computer and Information Science, Springer, vol. 413, pp. 120–135 (2013)
20. Swartout, W., Gratch, J., Hill, R., Hovy, E., Marsella, S., Rickel, J.: Toward virtual humans. AI Mag. **27**(2), 96–108 (2006)
21. Arafa, Y., Mamdani, A.: Virtual personal service assistants: real-time characters with artificial hearts. In: Proceedings of 1999 IEEE International Conference on Systems, Man, and Cybernetics, Tokyo, vol. 1, pp. 762–767 (1999)
22. Matsuyama, M., Bhardwaj, A., Zhao, R., Romero, O., Akoju, S., Cassell, J.: Socially-aware animated intelligent personal assistant agent. In: Proceedings of the 17th Annual SIGDIAL Meeting on Discourse and Dialogue, September 2016
23. Reina, A., Salvaro, M., Francesca, G., Garattoni, L., Pinciroli, C., Dorigo, M., Birattari, M.: Augmented reality for robots: Virtual sensing technology applied to a swarm of e-pucks. In: Proceedings of NASA/ESA Conference on Adaptive Hardware and Systems (AHS), Montreal, pp. 1–6 (2015)
24. Dragone, M., Duffy, B., O'Hare, G.: Social interaction between robots, avatars & humans. In: Proceedings of IEEE International Workshop on Robot and Human Interactive Communication, pp. 24–29 (2005)
25. Forland, E., Russa, G.: Virtual humans vs. anthropomorphic robots for education: how can they work together? In: Proceedings of ASEE/IEEE Frontiers in Education Conference, pp. S3G (2005)
26. Rahman, S., Ikeura, R.: Weight-prediction-based predictive optimal position and force controls of a power assist robotic system for object manipulation. IEEE Trans. Industrial Electron. **63**(9), 5964–5975 (2016)
27. Rahman, S., Ikeura, R.: Cognition-based control and optimization algorithms for optimizing human-robot interactions in power assisted object manipulation. J. Inf. Sci. Eng. **32**(5), 1325–1344 (2016)
28. Rahman, S., Sadr, B., Wang, Y.: Trust-based optimal subtask allocation and model predictive control for human-robot collaborative assembly in manufacturing. In: Proceedings of ASME Dynamic Systems and Controls Conference, Columbus, Ohio, October 28–30 (2015). Paper No. DSCC2015-9850, pp. V002T32A004
29. Rahman, S., Wang, Y., Walker, I.D., Mears, L., Pak, R., Remy, S.: Trust-based compliant robot-human handovers of payloads in collaborative assembly in flexible manufacturing. In: Proceedings of the 12th IEEE International Conference on Automation Science and Engineering (IEEE CASE 2016), Texas, USA, pp. 355–360. August 21–24 (2016)

30. Chipalkatty, R., Droge, G., Egerstedt, M.: Less is more: mixed-initiative model-predictive control with human inputs. IEEE Trans. Robot. **29**(3), 695–703 (2013)
31. Jones, A., Moulin, C., Barthès, J.P., Lenne, D., Kendira, A., Gidel, T.: Personal assistant agents and multi-agent middleware for CSCW. In: Proceedings of the 2012 IEEE 16th International Conference on Computer Supported Cooperative Work in Design (CSCWD), Wuhan, pp. 72–79 (2012)
32. Wahaishi, A.M., Aburukba, R.O.: An agent-based personal assistant for exam scheduling. In: Proceedings of 2013 World Congress on Computer and Information Technology (WCCIT), Sousse, pp. 1–6 (2013)
33. Sugawara, K., Manabe, Y., Fujita, S.: Mobile symbiotic interaction between a user and a personal assistant agent. In: Proceedings of 2012 IEEE 11th International Conference on Cognitive Informatics and Cognitive Computing, Kyoto, pp. 341–345 (2012)
34. Czibula, G., Guran, A.M., Czibula, I.G., Cojocar, G.S.: IPA—An intelligent personal assistant agent for task performance support. In: Proceedings of 2009 IEEE 5th International Conference on Intelligent Computer Communication and Processing, Cluj-Napoca, pp. 31–34 (2009)
35. Wong, W.S., Aghvami, H., Wolak, S.J.: Context-aware personal assistant agent multi-agent system. In: Proceedings of 2008 IEEE 19th International Symposium on Personal, Indoor and Mobile Radio Communications, Cannes, pp. 1–4 (2008)
36. Bush, J., Irvine, J., Dunlop, J.: Personal assistant agent and content manager for ubiquitous services. In: Proceedings of 2006 3rd International Symposium on Wireless Communication Systems, Valencia, pp. 169–173 (2006)
37. Blake, M.B.: Personal learning assistant agents in the business process domain. In: Proceedings of IEEE International Conference on Advanced Learning Technologies (ICALT 2006), Kerkrade, pp. 1117–1118 (2006)
38. Ma, C., Feng, J., Yang, Z., Wu, Q.: Agent-based personal article citation assistant. In: Proceedings of IEEE/WIC/ACM International Conference on Intelligent Agent Technology, pp. 702–705 (2005)
39. Nack, L., Roor, R., Karg, M., Kirsch, A., Birth, O., Leibe, S., Strassberger, M.: Acquisition and use of mobility habits for personal assistants. In: Proceedings of 2015 IEEE 18th International Conference on Intelligent Transportation Systems, Las Palmas, pp. 1500–1505 (2015)
40. Jalaliniya, S., Pederson, T.: Designing wearable personal assistants for surgeons: an egocentric approach. IEEE Pervasive Comput. **14**(3), 22–31, July-Sept. 2015
41. Sansen, H., Torres, M., Chollet, G., Glackin, C., Delacretaz, D., Boudy, J., Badii, A., Schlogl, S.: The Roberta IRONSIDE project: a dialog capable humanoid personal assistant in a wheelchair for dependent persons. In: Proceedings of 2016 2nd International Conference on Advanced Technologies for Signal and Image Processing (ATSIP), Monastir, pp. 381–386 (2016)
42. Chang, B., Tsai, H., Guo, C., Chen, C.: Remote cloud data center backup using HBase and Cassandra with user-friendly GUI. In: Proceedings of IEEE International Conference on Consumer Electronics, Taipei, pp. 420–421 (2015)
43. Kumar, N., Kumar, A., Giri, S.: Design and implementation of three phase commit protocol (3PC) directory structure through Remote Procedure Call (RPC) application. In: Proceedings of International Conference on Information Communication and Embedded Systems, pp. 1–5 (2014)
44. Rodriguez, I., Astigarraga, A., Jauregi, E., Ruiz, T., Lazkano, E.: Humanizing NAO robot teleoperation using ROS. In: Proceedings of 14th IEEE-RAS International Conference on Humanoid Robots, pp. 179–186 (2014)
45. Lee, J., Moray, N.: Trust, self-confidence, and operators' adaptation to automation. Int. J. Hum Comput Stud. **40**, 153–184 (1994)

Chapter 10
Emotion Detection and Regulation from Personal Assistant Robot in Smart Environment

José Carlos Castillo, Álvaro Castro-González, Fernándo Alonso-Martín, Antonio Fernández-Caballero and Miguel Ángel Salichs

Abstract This paper introduces a proposal for integrating personal assistant robots with social capacities in smart environments. The personal robot will be a fundamental element for the detection and healthy regulation of the affect of the environment's inhabitants. A full description of the main features of the proposed personal assistant robot are introduced. Also, the multi-modal emotion detection and emotion regulation modules are fully described. Machine learning techniques are employed for emotion recognition from voice and images and both outputs are merged to achieve the detected emotion.

10.1 Introduction

Smart environments evolve from ubiquitous computing following the idea of "a physical world that is richly and invisibly interwoven with sensors, actuators, displays, and computational elements, embedded seamlessly in the everyday objects of our lives, and connected through a continuous network" [1]. Smart environments are composed of several heterogeneous sensors placed throughout the environment, thus providing great amounts of data. Scalable and flexible platforms integrate such devices and provide applications with the necessary interfaces to interoperate with the information coming from the available resources. The main goal of a smart environment is using this information to achieve a more comfortable life for its inhabitants.

Some recent previous works of our research teams hold the objective to achieve emotion detection and regulation in smart environments through the incorporation of

J.C. Castillo (✉) · Á. Castro-González · F. Alonso-Martín · M.Á. Salichs
Department of Systems Engineering and Automatic, University Carlos III of Madrid, Getafe, Spain
e-mail: jocastil@ing.uc3m.es

A. Fernández-Caballero
Departamento de Sistemas Informáticos, Universidad de Castilla-La Mancha, Ciudad Real, Spain
e-mail: Antonio.Fdez@uclm.es

A. Costa et al. (eds.), *Personal Assistants: Emerging Computational Technologies*, Intelligent Systems Reference Library 132, DOI 10.1007/978-3-319-62530-0_10

several sensing and actuation technologies. The ultimate aim is to maintain a healthy affective state of the subject. In our opinion, the inclusion of a robotic platform in such smart environments opens new possibilities for perceiving the inhabitant's emotional state and properly acting on his/her mood. In particular, social robots and humans tend to establish affective bonds that could be exploited in a smart environment. Therefore, these robotic platforms are not only a mere set of mobile sensors and actuators as robot companions but also a way to offer easy and amiable interaction between humans and the smart environment.

This paper introduces a proposal to incorporate a personal assistant robot with social skills into a smart environment for the sake of complementing intelligent affective detection and regulation strategies. According to the European Commission, "personal assistant robots would be able to learn new skills and tasks in an active open-ended way and to grow in constant interaction and co-operation with humans" [2]. Here, it is our intention to establish the foundations for designing social personal assistant robots as an outstanding part to enhance the capabilities of smart environments.

The rest of the proposal is described next. Section 10.2 introduces the description of the proposed personal assistant robot. This section shows the main features that the social robot has to build-in, that is, it has to be mobile, it has to wear *ears* and *eyes*, and it has to incorporate expressive capabilities. Afterwards, in order to achieve affective detection, Sect. 10.3 describes the main features of the multi-modal emotion detection module included in the social robot. Emotion detection is done in this proposal through applying machine learning techniques to both voice and video streams, analysing the sensory data separately and then merging both outcomes. Next, the proposed affective regulation module is described in Sect. 10.4. Lastly, our conclusions are summarised in Sect. 10.5.

10.2 The Personal Assistant Robot

As described before, in previous works we have been designing smart environments endowed with perception and action capabilities to detect and regulate the user's emotional state [3, 6]. However, this approach presents several drawbacks and limitations. Firstly, the environment has to be altered resulting in high economic cost of installation and maintenance. Besides, the installed devices/sensors are usually perceived by the user as moderately to very intrusive; and therefore the smart system could be rejected. Finally, the location of the sensors is static. Thus, the monitoring elements are located far from the user (walls or ceiling) most of the time, lowering the quality of the data captured.

The use of a social robot mitigates these problems. Social robots are intended to interact with people following established behavioural norms [7]. These robots coexist in daily environments (e.g. homes, schools, hospitals or museums) helping humans to perform particular tasks, assisting patients with their therapies, or just accompanying people. This kind of activities requires the robot appearance to be appealing and friendly, unlike traditional industrial robots for instance. Consequently social robots need to be carefully designed. Traditionally, these robots present an external look similar to animals (dog robot [8], cat robot [9], a seal robot [10]), cartoons (big eyes, or round shapes), or even a mix of both, such us the little, furry DragonBot [11].

A social robot's friendly appearance will for sure help to be accepted by the users and emotional bonds, similar to those existing with a pet or a friend, will arise. The inclusion of head, face and arms in the robot benefits perception of a living entity or an animal. In addition, all sensors are located on board, avoiding the need of a physical set-up and the modification of the environment. Moreover, this implies that the robot is able to move around the environment to approach the user, or to move to another room. Therefore, its is mandatory that the social robot integrates a mobile platform. Most of the human-robot interactions with social robots are usually conducted is short distances (around few meters), similar to human-human interactions. This proximity between the user and the robot ensures the acquisition of a high quality data from the embarked sensors.

Another important aspect is the robot's size. This will depend on the users it is intended to interact with. In usual daily living environments, the robot would live with adults who are able to easily communicate with the robot in a natural manner. A small robot would cause users to bend when communicating, and a big robot could be overwhelming. In consequence, considering the average height of adult men and women, we believe that the robot's height should range between 1.5 and 1.7 m.

In the next subsections, the required hardware (sensors and actuators) that are needed by the robotic platform to successfully operate in a smart environment is described.

10.2.1 A Mobile Social Robot

As already stated, the social robot needs to autonomously move around to approach, follow, or search for users. To date, the most reliable technology to achieve this functionality is a mobile platform based on wheels. In order to allow a safe navigation, the robot needs a map of the environment for detecting obstacles and calculating the trajectory from one point to another. These tasks require that the robot gets data to build the map and perform obstacle detection.

A clever combination is to install a 2D LIDAR in the base and a 3D depth camera. The LIDAR device provides accurate data to build the map and to navigate. Nevertheless, it only provides 2D data that could be insufficient to detect obstacles like a table or a stretcher. This type of barriers are easily detected by a 3D camera, like the well-known RGB-D Kinect sensor. The merged data from these two types of devices, together with the precise control of the motorised wheels, results in a safe and reliable navigation of our mobile platform.

10.2.2 A Social Robot with Ears and Eyes

Based on previous works (e.g. [12, 13]), we want to use a mobile, social robot as a smart sensor to estimate the user's emotional state or mood. We will use two sources of information mainly, the user's voice and image. This is why our social robot needs "ears" and "eyes".

In order to detect users utterance, we suggest to endow the robot with an array of directional microphones that provide high quality sound, even if the source is not very close, but at the cost of a narrow operational angle. This is why we need a ring of micros around the robot. As an extra benefit, the ring of microphones allows the localisation of the audio source, which results very interesting during human-robot interaction.

In the audio domain, the emotional content of the signal can provide helpful information to asses the user's state—for instance, the tone of voice of a person about to cry is completely different to a happy person's one. In relation with the robot's "eyes", the user's face is analysed to assess the current emotion [16, 17].

10.2.3 A Social Robot with Expressive Capabilities

In this paper, our intention is to present a system that combines a smart environment with a social robot to modify the user's emotional state. Therefore, the robot needs to have expressive capabilities to impact on the user's affective state. The way the robot moves could affect the user's perception of the robot. In case the robot moves with abrupt movements, the robot could be perceived as dangerous and the user would feel agitated. On the contrary, if the robot performs soft, smooth movements, it could contribute to calm the user.

Many social robots are endowed with screens (for example in the eyes, the face, or the body [14, 15]). The dominant colour displayed on the screens can change according to the desired effect on the user, as described later on in the section dedicated to emotion regulation. Also the multimedia content shown can help to communicate an emotion: high-valence and low-arousal images similar to those labelled in the International Affective Picture System database [18] help to calm down a person, or emoticons help to perceive the robot's artificial emotional state (like smiling or sad faces). Similarly to the intention followed with screens, some recent robots are endowed with a projector to display images on the environment. Due to the impact of the projections on the surroundings, this sounds like a promising method to guide the user's mood. Besides, colour LEDs can be placed in different parts of the robot, e.g. chicks, mouth, or base. The colour displayed in these elements would depend on the user requirements.

Finally, other crucial robotic elements capable of altering the human mood are non-verbal sounds. Different sounds full of emotional content can be synthesized or reproduced by the robot. Most of these elements, both sensors and actuators, are

Fig. 10.1 Maggie (*left*) and MBot (*right*), the mobile social robots from the Robotics Lab at Universidad Carlos III de Madrid

present in the social robots from the Robotics Lab at Carlos III University of Madrid (Fig. 10.1). Additionally, these robots are a good example of mobile platforms with a friendly external appearance and an appropriate size.

10.3 The Multi-modal Emotion Detection Module

The emotion detection module combines analysis in two different domains. On the one hand, it uses voice analysis techniques to assess emotions in the user's speech. Computer vision complement the audio-based approach by adding face detection and recognition capabilities that are also able to detect emotions in video sequences. Finally, a bayesian decision rule is in charge of merging the outputs from both analyses, generating a unique result. The combination of both approaches is described in Fig. 10.2.

In the figure, GEVA is the component in charge of emotion detection through voice analysis while GEFA is in charge of emotion detection through video (facial expressions) analysis. Therefore, the input to GEVA is the user's voice, where features are extracted and a couple of classifiers obtain one of the basic emotions "neutral", "happiness", "sadness" and "surprise". Complementary, GEFA uses a couple of tools (as described in Sect. 10.3.2) to obtain the same basic emotions. Lastly, there is a decision rule that uses the output of both components to get a final consensus basic emotion as detailed in Sect. 10.3.3.

Four basic emotions, "neutral", "happiness", "sadness" and "surprise", have been chosen for two main reasons: The first one is that these basic emotions can be represented very easily in a classical circumplex model of affect [19]. This model suggests that emotions are distributed in a two-dimensional circular space, containing arousal (activation) and valence (pleasantness) dimensions. Arousal represents the vertical axis and valence expresses the horizontal axis, while the centre of the circle means a

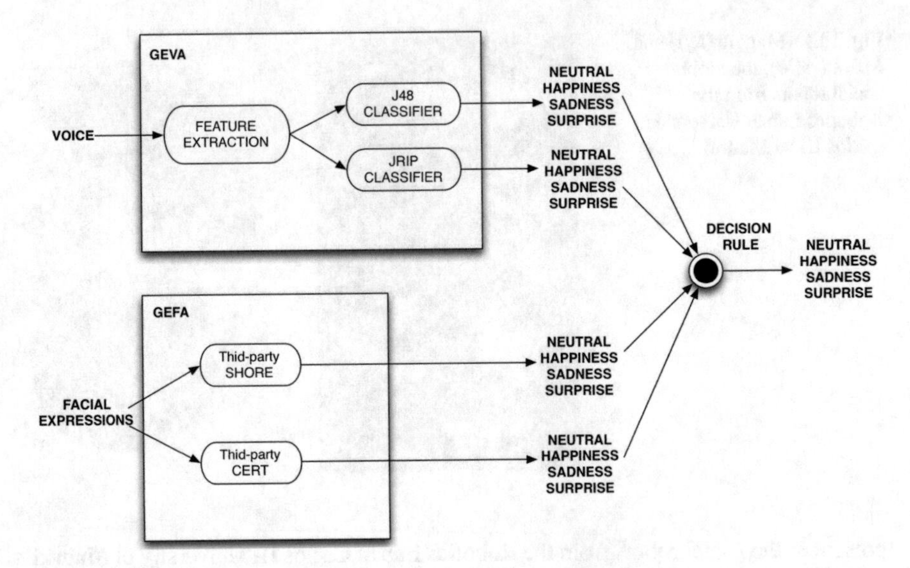

Fig. 10.2 General scheme of the multimodal emotion detection module

Fig. 10.3 Circumplex model of affect including the four basic emotions considered

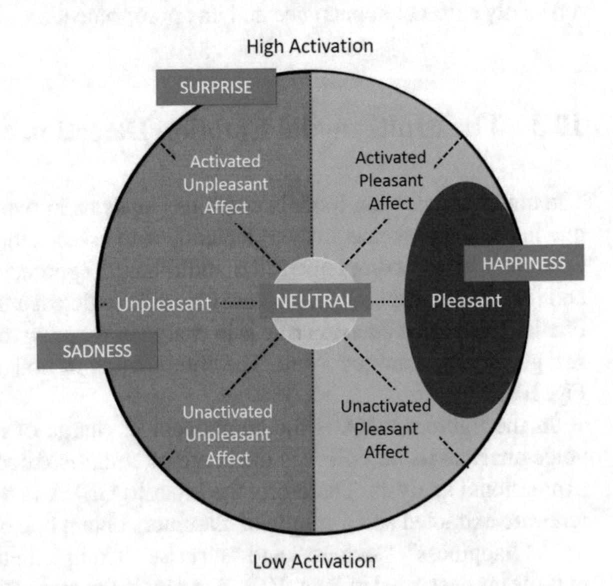

neutral valence and a medium level of arousal. In this model, emotional states can be represented at any level of valence and arousal, or at a neutral level of one or both of these factors. Figure 10.3 offers a circumplex model of affect particularised for the four emotions considered in this proposal.

In second place, the pathways to regulate the chosen emotions are also well-known in Psychology. In fact, empirical evidence shows that the two emotional dimensions of arousal and valence are not independent of each other [20]. Rather, they form a V ("boomerang")-shaped function, with the unpleasant stimuli tending to be more highly arousing than pleasant stimuli, and both pleasant and unpleasant stimuli being more arousing than neutral stimuli. Considerable evidence supports this "boomerang" shape in averaged data on arousal and valence ratings of people's reactions to affective visual stimuli. In Fig. 10.3 you may also observe that the two axis split the circumference into four parts denominated *Activated Unpleasant Affect*, *Unactivated Unpleasant Affect*, *Unactivated Pleasant Affect* and *Activated Pleasant Affect*.

10.3.1 Emotion Detection Through Voice Analysis

Changes in emotions can be detected by analysing the interlocutor's voice tone [21]. This proposal exploits the capabilities of a multi-domain audio-based emotion detection approach named GEVA [22, 23]. It extracts some relevant features from the input signals that are sent to a classifier that obtains the emotion as an output.

10.3.1.1 Feature Extraction

In this first phase of the GEVA component, the analysis is performed in three domains: (i) time domain, in which the analysis is directly performed in the analogue signal; (ii) frequency domain, in which the Fast Fourier Transform is applied to the input signal; and (iii) time-frequency domain, in which the analysis is performed after applying the Discrete Haar Wavelet transform to the input data. Features are extracted in these three domains using Chuck,[1] a programming language for real-time sound synthesis and sound wave analysis. The features extracted are the following ones:

- **Pitch**: Sound frequency as perceived by the human ear.
- **Flux**: How big the sound variations are in the frequency domain. Values close to zero indicate small differences whilst values near one imply that there are important variations in the frequency domain.
- **Rolloff-95**: Frequency value at which 95% of the signal energy is already contained.
- **Centroid**: Median of the signal spectrum. That is, the frequency the signal approaches the most. This frequency is related to the tone of a sound (timber).
- **Zero-crossing rate**: Number of times that the signal crosses the zero (x axis). This is useful to distinguish between background noise and voice since the former tend to cross the axis more frequently than the latter.

[1] Chuck website: http://chuck.cs.princeton.edu/uana/.

- **Signal-to-noise ratio (SNR)**: Voice signal volume with respect to the background noise.
- **The communicative rhythm**: This feature is defined as the number of words pronounced per minute. This is useful when trying to distinguish emotions since each of them have characteristic communication rhythms.

10.3.1.2 Classification

The features described in the previous phase characterise the main parameters of the acquired sound wave. Next, these are used to train a classifier that obtains the detected emotion. GEVA implements an off-line universal classifier, trained with no constraint regarding the type of user. Therefore, the dataset for training contains sentences from a wide range of users of different age, language and gender. The dataset contains samples of tagged locutions for the four target emotions: "happiness", "sadness", "surprise" and "neutral". The training dataset is composed from several sources:

- Voice examples from the developers simulating emotions.
- Interviews with colleagues asked to fake emotions.
- Real or spontaneous interactions between the robot and colleagues.
- Interviews obtained from the Internet.
- TV shows from the Internet.
- Audiobooks from the Internet.
- Databases with a tagged voice corpus:
  - Emotional Prosody Speech Database (LDC): with 15 types of emotions [24].
  - Berlin Emotional Speech Database (EmoDB): with seven types of emotions [25].
  - Danish Emotional Speech Corpus (HUMANAINE): with five types of emotions [26].
  - FAU Aibo Emotion Corpus: 8.9 h of spontaneous voice recordings from 51 children, with 11 types of emotions[2] [27].

In order to find the best-suited classifier for emotion detection we have used the software library Weka [28], which integrates more than one hundred machine learning techniques.

Using our linguistic corpus and the voice feature extraction module, a file containing about 500 emotion-tagged locutions with training patterns for Weka is built. Using cross-validation over the dataset, two algorithms are selected: a decision tree-based algorithm, J48, and a decision rule-based one, JRIP, with a 80.51% and 81, 15% of success rate, respectively (see [23] for more details).

[2]Fau website: http://www5.cs.fau.de/de/mitarbeiter/steidl-stefan/fau-aibo-emotion-corpus.

10.3.2 Emotion Detection Through Video Analysis

The literature offers a number of techniques for emotion detection from facial expressions [16, 17]. Most of these techniques follow the same steps: (i) face detection in the image flow; (ii) facial feature extraction, such as distance between eyes or mouth shape, among others; and (iii) emotion classification using the previous features and several learning techniques.

Currently, there is a number of algorithms for user face detection and tracking in multiple applications. Usually, these techniques focus on the detection of frontal faces, identifying candidate regions for characteristic components of a face such as eyes, nose or mouth. Among the most widespread techniques we may find the Viola and Jones algorithm [29], although there are other trends such as the ones based on machine learning, e.g. neural networks [30] or support vector machines (SVM) [31]. Besides, well-known image processing libraries such as OpenCV[3] also include other wide-spread algorithms for face detection.

After locating a face in the image, it is necessary to extract features to simplify the classifiers' work. For this purpose, there are approaches based on interest points and their geometric relationships (local approximation) [32, 33]. Also, some works deal with this problem through representing the face as a whole unit, e.g. placing a 3D mesh over the detected face, calculating the differences between the current detection and a target one (holistic approach) [34, 36]. These two approaches can be mixed together by using the interest points to determine the initial position of the mesh [37].

There are works that verse about the possibility of having a universal facial expressions classification for emotion recognition, taking into account gender, age and culture [38, 39]. This is related to the different intensities when expressing emotions which make the recognition problem more challenging when the intensity is low, as facial variations are more subtle.

In conjunction with the advances in the literature, some works propose off-the-shelf tools for face detection, feature extraction and emotion classification. An example is CERT [40], a visual tool that allows classifying six emotions in real time as well as 16 action units (AU) from the Facial Action Coding System [41]. Feature extraction follows a local approximation-based approach and AUs are classified using a SVM with results ranging between 80 and 90% for on-line emotion recognition. CERT outputs the intensity value for each detected emotion. In this case, the outputs are: fun, joy, smile detector (these three are grouped in one set as happiness), disgust and sadness (grouped as sadness), surprise, neutral, fear and anger (these last two are not considered in our approach).

There is another interesting tool, SHORE [42, 43], for face detection and emotion recognition in challenging environments (e.g. lightning changes). This software implements a holistic approach able to track the position and orientation of user faces although the detection performance is better when dealing with front views. SHORE provides the intensity values of the following emotional states: happiness, sadness,

[3]OpenCV website: http://opencv.org.

surprise and anger. Again, this last emotion is not considered in our proposal. Moreover, if none of these emotional states has an intensity greater than 50%, then it is assumed that the emotional state of the user is neutral.

These two tools, CERT and SHORE have been used to build the GEFA module. Using these two tools, GEFA is able to operate at a maximum interaction distance of 4.5 m.

10.3.3 Integration of GEVA and GEFA

The outputs of the GEVA and GEFA emotion detection components need to be merged in order to get a unified outcome. Figure 10.4 shows a scheme of the process to determine the user emotion. A previous work [23] demonstrates that the accuracy of the visual emotion detection system drops when users are talking, as the image libraries for training are mainly composed of non-speaking face images. Therefore, when a user is quiet, the output of GEFA is considered and, otherwise, the information coming from GEVA is the one taken into account. These outputs may be seen as independent, since they occur at different time instants. Nevertheless, emotions do not occur instantaneously, so a decision rule can be useful to combine the information from both emotion detection modules, generating a single output with lower temporal resolution.

Once the success rate of each module (with two classifiers each) is estimated, a decision rule is applied in order to create a unified output. Prior to that, it is necessary to determine the *confidence degree* of the output (the detected emotion) for each

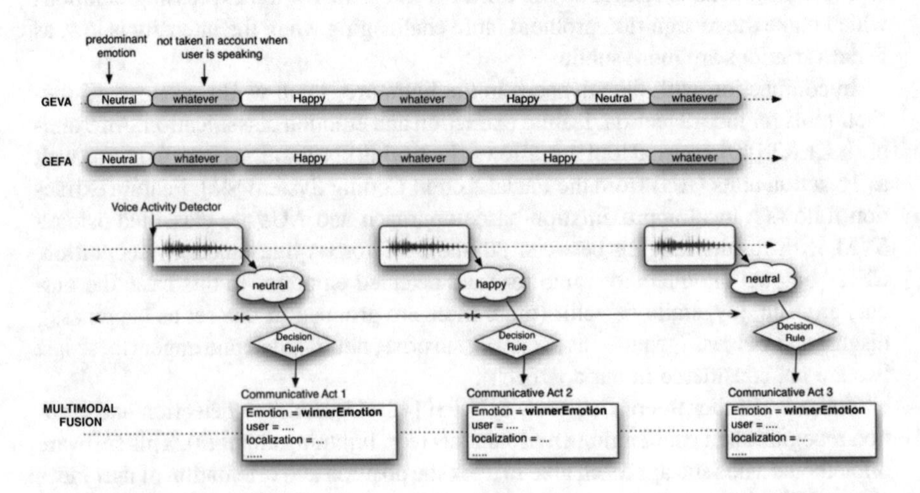

Fig. 10.4 Graphical representation of the proposed system. Outputs from GEVA and GEFA are fused using the decision rule

classifier. In this work, the *confidence degree* of each output is calculated using the Bayes theorem and the confusion matrices. The decision rule used is as follows:

Decision rule: The easiest way of determining the most probable state of the system is to find the state $\hat{s}$ that has the highest probability among all classifiers given by:

$$\hat{s} \in \mathbf{S}, \hat{C} \in \mathbf{C} \mid \forall s_i \in \mathbf{S}, \forall C \in \mathbf{C}\ ,\ p(S = s_i|S_C) \leq p(S = \hat{s}|S_{\hat{C}}). \tag{10.1}$$

where the current state of the system is S and the list of states is $\mathbf{S} = \{s_1, \ldots, s_n\}$ where $n \in \mathbb{N}$, and $\mathbf{C}$ is a collection of m classifiers $\mathbf{C} = C_1, \ldots, C_m, m \in \mathbb{N}$.

This decision rule allows fusing the four outputs of the integrated classifiers in a single one with a higher accuracy with respect to the isolated original ones. Details about the mathematics to estimate the conditional probabilities that result in the decision rule are available in a previous paper [23].

10.4 The Emotion Regulation Module

The aim of the emotion regulation module is to provide the best-suited conditions of music performance, as well as colour/light to attain the desired affect in the person interacting with the personal assistant robot. We propose to simultaneously use multiple induction techniques to regulate the user' mood in his/her smart environments. These techniques are fully incorporated both in the social robot and in the smart ambient. Taking advantage of the actuators embarked in the robot's body as well as the ones in the ambient, music is performed by means of an box that plays background music. And, colours are automatically displayed in the robot's parts as well as in the environment in a smart manner.

The different emotions are detected/regulated through the combination of activation/pleasantness values and position in the orthogonal dimensions.

- Neutral: medium activation/medium pleasantness; undefined orthogonal dimension.
- Sadness: medium activation/low pleasantness; Unactivated Unpleasant Affect.
- Happiness: medium activation/high pleasantness; Activated Pleasant Affect.
- Surprise: high activation; medium pleasantness; Activated Unpleasant Affect.

10.4.1 Musical Emotion Regulation

A research work states that the most potent and frequently studied musical cues for emotion elicitation are mode, tempo, dynamics, articulation, timbre, and phrasing [44]. In fact, musical parameters such as tempo or mode are inherent properties of

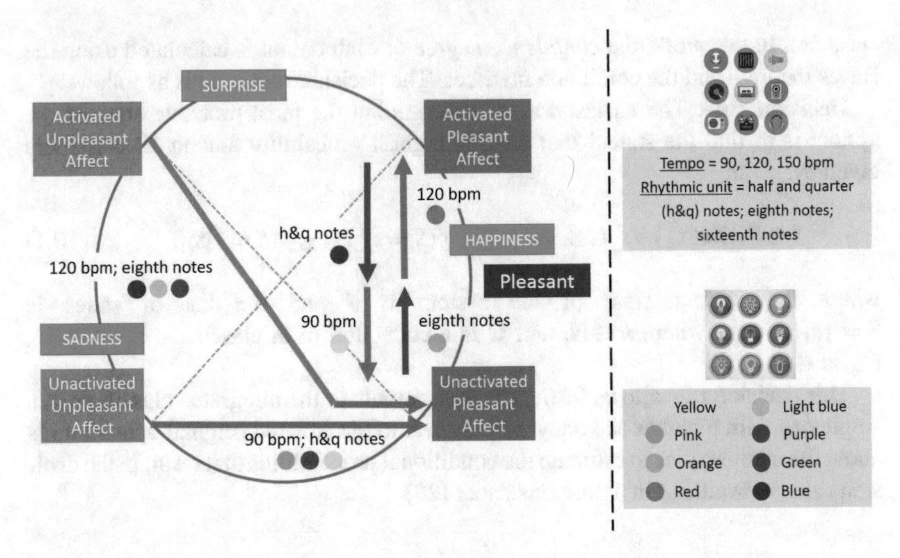

Fig. 10.5 Circumplex model for emotion regulation from personal assistant robot

musical structure [45]. These characteristics are important, as it has been shown that they influence listeners' emotions.

Musical emotion regulation focuses in our case on note value, an important musical cue related to rhythm. The two musical components of note value used are tempo and rhythmic unit. Beyond the effect of mode, tempo was found to modulate the emotional ratings, with faster tempo being more associated with the emotions of anger and happiness as opposed to slow tempo, which induces stronger feeling of sadness and serenity. This suggests that the tempo modulates the arousal value of the emotions [46].

Figure 10.5 shows the values of tempo and rhythmic unit that best provide the desired transitions from negative to positive affect in accordance with some of our prior experimental works [47, 48]. This figure is also based in the expert knowledge on the "boomerang" shape that psychologists and neuroscientists have described during the two last decades. At this point, it is important to clarify that the intervention proposed in this work is for situations where a negative affective state is unhealthy and a transition to a positive emotional state is necessary.

Thus, depending on the starting position of the emotion detected in the orthogonal dimensions, we have:

- *Activated Unpleasant Affect* is an unhealthy affective state, so that a transition to a healthy state is required. In accordance with psychological studies, firstly the activation level has to be dropped. In this case (see brown line in Fig. 10.5), a musical piece mostly based on eighth notes and a tempo of 120 beats per minute (bpm) is played. Once, the activation is below a given value, another musical piece is played (90 bpm, and half and quarter notes), which is shown with a green arrow. Notice that a complete and varied battery of musical pieces is available.

- *Unactivated Unpleasant Affect* is also an unhealthy emotional state. In this case, it is not necessary to deactivate the patient. Just as in the previous case, a musical piece is played in 90 bpm, and half and quarter notes.
- *Unactivated Unpleasant Affect* is a health affective state. However, in this proposal we will be trying to maintain the patient as "Pleasant" as possible. This is why, it is foreseen that the affective state will balance between *Unactivated Unpleasant Affect* and *Activated Pleasant Affect*, depending on the reaction of the monitored person. In order to pass to *Activated Pleasant Affect*, two steps are necessary. In the first one, a musical piece based on eighth notes is played, and in the second one it is a musical piece played at 120 bpm.
- The *Activated Pleasant Affect* case is similar to the previous one. Here, firstly a musical piece based on half and quarter notes is played, and then a musical piece is played at 90 bpm.

10.4.2 Colour/Light-Based Emotion Regulation

Light is one of the principal environmental factors that influence on a user. Therefore, its affective impact has been widely researched and reported. Exposure to bright light and luminance distribution affect our mood. This is why, many researchers believe that light can be used to alter/improve emotional states. Colour is another factor that is constantly present in any environment. Moreover, the evidences of the influence of colour upon human emotions has been discussed in a number of previous publications from our research team (e.g. [49, 50]). Many authors highlight that certain colours are highly influencing for emotion elicitation and regulation.

Therefore, colour emotion regulation bases here on the combination of colour and light due to their double influence on emotion regulation. Also, in Fig. 10.5 you can appreciate the colours that best provide the desired transitions from negative to positive mood in accordance with a prior work [51]. Again, depending on the starting position of the emotion detected in the orthogonal dimension, we have:

- *Activated Unpleasant Affect* has to transit to *Unactivated Pleasant Affect*. This is achieved in two steps. Firstly, colour/light is changed to Blue, Light blue or Purple, and, then to Pink, Green or Yellow. The order in these lists corresponds to the velocity at which these colours provide the desired effect. Obviously, this has also to be personalised in accordance with the user's reactions.
- *Unactivated Unpleasant Affect* transits directly to *Unactivated Pleasant Affect* with colour/light Pink, Green or Yellow.
- Remember that *Unactivated Unpleasant Affect* is a healthy affective state, but we try to maintain the patient as "Pleasant" as possible. This is why, it is foreseen that the affective state will balance between *Unactivated Unpleasant Affect* and *Activated Pleasant Affect*. In order to transit to *Activated Pleasant Affect*, newly two steps are necessary. In the first one, colour/light is changed to Orange, and in the second one colour/light is changed to Red.

- The *Activated Pleasant Affect* case is similar to the previous one. Here, firstly colour/light is changed to Blue, and in the second one colour/light is changed to Yellow.

10.5 Conclusions

This work offers a proposal for a smart environment for emotion regulation including the capabilities of a personal assistant robot. Although traditionally sensors are placed throughout the environment, we propose an innovative approach where several sensing devices are embarked in a social robot. This robot is in charge of data acquisition and processing. Additionally, the environment and the robot are employed to influence the user's mood.

We propose the main features a robotic platform ought to incorporate such as a mobile base with sensors for navigation. The proposed robot needs to be endowed with screens, colour LEDs, audio devices to allow high expressibility to produce the desired effect on the user.

One important aspect of the proposal is the ability to detect emotions. This is achieved by equipping the robot with cameras and microphones which output is fused using a Bayes-based decision rule. Four basic emotions (neutral, happiness, sadness, and surprise) are detected using the analysis modules, which rely on machine learning techniques based on computer vision and audio features classification. On top of that, an approach for emotion regulation using music and colour has been presented, showing how tempo and rhythmic units have to be modulated, in the case of music, and what colours provide the transitions between moods.

Acknowledgements This work was partially supported by Spanish Ministerio de Economía y Competitividad/European Regional Development Fund under TIN2015-72931-EXP and DPI2016-80894-R grants. The research leading to this work has also received funding from the projects: Development of social robots to help seniors with cognitive impairment (ROBSEN), funded by the Ministerio de Economia y Competitividad/European Regional Development Fund; and RoboCity2030-III-CM, funded by Comunidad de Madrid and co-funded by Structural Funds of the EU.

References

1. Weiser, M., Gold, R., Brown, J.S.: The origins of ubiquitous computing research at PARC in the late 1980s. IBM Syst. J. **38**(4), 693–696 (1999)
2. European Commission.: Beyond Robotics (RO) Proactive Initiative (2006). https://cordis.europa.eu/ist/fet/ro-in.htm
3. Castillo, J.C., Castro-González, Á., Fernández-Caballero, A., Latorre, J.M., Pastor, J.M., Fernández-Sotos, A., Salichs, M.A.: Software architecture for smart emotion recognition and regulation of the ageing adult. Cogn. Comput. **8**(2), 357–367 (2016)

4. Fernández-Caballero, A., Martínez-Rodrigo, A., Pastor, J.M., Castillo, J.C., Lozano-Monasor, E., López, M.T., Zangróniz, R., Latorre, J.M., Fernández-Sotos, A.: Smart environment architecture for emotion recognition and regulation. J. Biomed. Inf. **64**, 55–73 (2016)
5. Castillo, J.C., Fernández-Caballero, A., Castro-González, Á., Salichs, M.A., López, M.T.: A framework for recognizing and regulating emotions in the elderly. Ambient Assisted Living and Daily Activities, pp. 320–327 (2014)
6. Fernández-Caballero, A., Latorre, J.M., Pastor, J.M., Fernández-Sotos, A.: Improvement of the elderly quality of life and care through smart emotion regulation. Ambient Assisted Living and Daily Activities, pp. 348–355 (2014)
7. Bartneck, C., Forlizzi, J.: A design-centred framework for social human-robot interaction. In: 13th IEEE International Workshop on Robot and Human Interactive Communication, pp. 591–594 (2004)
8. Moon, Y.E.: Sony AIBO: the world's first entertainment robot. Harvard Business School Case 502-010 (2001)
9. van Breemen, A., Yan, X., Meerbeek, B.: iCat: an animated user-interface robot with personality. In: The Fourth International Joint Conference on Autonomous Agents and Multiagent Systems, pp. 143–144 (2005)
10. Shibata, T., Inoue, k., Irie, R.: Emotional robot for intelligent system-artificial emotional creature project. In: 5th IEEE International Workshop on Robot and Human Communication, pp. 466–471 (1996)
11. Setapen, A., Breazeal, C.: DragonBot: a platform for longitudinal cloud-HRI. Human-Robot Interaction (2012)
12. Jiang, M., Zhang, L.: Big data analytics as a service for affective humanoid service robots. Proc. Comput. Sci. **53**, 141–148 (2015)
13. Alvarez, M., Galan, R., Matia, F., Rodriguez-Losada, D., Jimenez, A.: An emotional model for a guide robot. IEEE Trans. Syst. Man Cybern.? Part A Syst. Hum. **40**(5), 982–992 (2010)
14. Pérula-Martínez, R., Salichs, E., Encinar, I.P., Castro-González, Á., Salichs, M.A.: Improving the expressiveness of a social robot through luminous devices. In: 10th ACM/IEEE International Conference on Human-Robot Interaction. Extended Abstracts, pp. 5–6 (2015)
15. Mirnig, N., Tan, Y.K., Chang, T.W., Chua, Y.W., Dung, T.A., Li, H., Tscheligi, M.: Screen feedback in human-robot interaction: how to enhance robot expressiveness. In: The 23rd IEEE International Symposium on Robot and Human Interactive Communication, pp. 224–230 (2014)
16. Pantic, M., Rothkrantz, L.: Automatic analysis of facial expressions: the state of the art. IEEE Trans. Pattern Anal. Mach. Intell. **22**, 1424–1445 (2000)
17. Khatri, N.N., Shah, Z.H., Patel, S.A.: Facial expression recognition: a survey. Int. J. Comput. Sci. Inf. Technol. **5**(1), 149–152 (2014)
18. Lang, P.J.: The emotion probe: studies of motivation and attention. Am. Psychol. **50**(5), 372–385 (1995)
19. Russell, J.A.: A circumplex model of affect. J. Pers. Soc. Psychol. **39**(6), 1161–1178 (1980)
20. Libkuman, T.M., Otani, H., Kern, R., Viger, S.G., Novak, N.: Multidimensional normative ratings for the international affective picture system. Behav. Res. Methods **39**, 326–334 (2007)
21. Cowie, R., Douglas-Cowie, E., Romano, A.: Changing emotional tone in dialogue and its prosodic correlates. In: ESCA Tutorial and Research Workshop on Dialogue and Prosody, pp. 41–46 (1999)
22. Alonso-Martin, F., Castro-González, A., Gorostiza, J., Salichs, M.A.: Multidomain voice activity detection during human-robot interaction. In: International Conference on Social Robotics, pp. 64–73 (2013)
23. Alonso-Martin, F., Malfaz, M., Sequeira, J., Gorostiza, J., Salichs, M.A.: A multimodal emotion detection system during human-robot interaction. Sensors **13**(11), 15549–15581 (2013)
24. Liberman, M., Davis, K., Grossman, M., Martey, N., Bell, J.: Emotional Prosody Speech and Transcripts. Linguistic Data Consortium, Philadelphia (2002)
25. Vlasenko, B., Schuller, B.: Combining frame and turn-level information for robust recognition of emotions within speech. Interspeech, pp. 27–31 (2007)

26. Schuller, B., Arsic, D.: Emotion recognition in the noise applying large acoustic feature sets. Speech Prosody, 276–289 (2006)
27. Steidl, S.: Automatic Classification of Emotion Related User States in Spontaneous Children's Speech, pp. 1–250. University of Erlangen, Logos-Verlag (2009)
28. Holmes, G., Donkin, A., Witten, I.: WEKA: a machine learning workbench. The IEEE Australian New Zealand Intelligent Information Systems Conference, pp. 357–361 (1994)
29. Viola, P., Jones, M.J.: Robust real-time face detection. Int. J. Comput. Vis. **57**(2), 137–154 (2004)
30. Rowley, H.A., Baluja, S., Kanade, T.: Neural network-based face detection. IEEE Trans. Pattern Anal. Mach. Intell. **20**(1), 23–38 (1998)
31. Osuna, E., Freund, R., Girosit, F.: Training support vector machines: an application to face detection. In: IEEE Computer Society Conference on Computer Vision and Pattern Recognition, pp. 130–136 (1997)
32. Kobayashi, H., Hara, F.: Facial interaction between animated 3d face robot and human beings. In: The IEEE International Conference on Systems, Man, and Cybernetics. Computational Cybernetics and Simulation 4, pp. 3732–3737
33. Padgett, C., Cottrell, G.: Representing face images for emotion classification. Advances in Neural Information Processing Systems 9 (1997)
34. Cootes, T., Edwards, G., Taylor, C.: Active appearance models. In: 5th European Conference on Computer Vision, pp. 484–498 (1998)
35. Terzopoulos, D., Waters, K.: Analysis and synthesis of facial image sequences using physical and anatomical models. IEEE Trans. Pattern Anal. Mach. Intell. **15**, 569–579 (1993)
36. Lucey, S., Matthews, I., Hu, C.: AAM derived face representations for robust facial action recognition. In: 7th International Conference on in Automatic Face and Gesture Recognition (2006)
37. Kearney, G., McKenzie, S.: Machine interpretation of emotion: design of a memory-based expert system for interpreting facial expressions in terms of signalled emotions. Cogn. Sci. **17**, 589–622 (1993)
38. Ekman, P., Friesen, W.: Constants across cultures in the face and emotion. J. Pers. Soc. Psychol. **17**(2), 124–129 (1971)
39. Russell, J., Dols, J.: The Psychology of Facial Expression. Cambridge University Press (1997)
40. Littlewort, G., Whitehill, J., Wu, T.-F., Butko, N., Ruvolo, P., Movellan, J., Bartlett, M.: The motion in emotion—a CERT based approach to the FERA emotion challenge. Face Gesture **2011**, 897–902 (2011)
41. Ekman, P., Friesen, W., Hager, J.: Facial action coding system: A technique for the measurement of facial movement. Number A Human Face. Consulting Psychologists Press, Palo Alto, USA (1978)
42. Wierzbicki, R.J., Tschoeppe, C., Ruf, T., Garbas, J.U.: EDIS-emotion-driven interactive systems. Int. SERIES Inf. Syst. Manag. Creat. Media **1**, 59–68 (2013)
43. Küblbeck, C., Ernst, A.: Face detection and tracking in video sequences using the modified census transformation. Image Vis. Comput. **24**, 564–572 (2006)
44. Gabrielsson, A., Lindstrom, E.: The role of structure in the musical expression of emotions. In: Theory, Research, and Applications, Handbook of Music and Emotion, pp. 367–400 (2010)
45. van der Zwaag, M.D., Westerink, J.L., van den Broek, E.L.: Emotional and psychophysiological responses to tempo, mode, and percussiveness. Musicae Sci. **15**(2), 250–269 (2011)
46. Trochidis, K., Bigand, E.: Investigation of the effect of mode and tempo on emotional responses to music using EEG power asymmetry. J. Psychophysiol. **27**(3), 142–147 (2013)
47. Fernández-Sotos, A., Fernández-Caballero, A., Latorre, J.M.: Influence of tempo and rhythmic unit in musical emotion regulation. Front. Comput. Neurosci. **10**, 80 (2016)
48. Fernández-Sotos, A., Fernández-Caballero, A., Latorre, J.M.: Elicitation of emotions through music: the influence of note value. In: Artificial Computation in Biology and Medicine, 488–497 (2014)
49. Sokolova, M.V., Fernández-Caballero, A., Ros, L., Fernández-Aguilar, L., Latorre, J.M.: Experimentation on emotion regulation with single-colored images. In: ICT-based Solutions in Real Life Situations, Ambient Assisted Living, pp. 265–276 (2015)

50. Sokolova, M.V., Fernández-Caballero, A.: A review on the role of color and light in affective computing. Appl. Sci. **5**(3), 275–293 (2015)
51. Ortiz-García-Cervigón, V., Sokolova, M.V., García-Muñoz, R., Fernández-Caballero, A.: LED strips for color- and illumination-based emotion regulation at home. In: ICT-based Solutions in Real Life Situations, Ambient Assisted Living, pp. 277–287 (2015)

Part VI
Ethic and Social Issues

Chapter 11
EDI for Consumers, Personal Assistants and Ambient Intelligence—The Right to Be Forgotten

Francisco Pacheco de Andrade, Teresa Coelho Moreira, Mikhail Bundin and Aleksei Martynov

Abstract Electronic Data Interchange has been around for more than 20 years but it is now on the verge of transition to the new environments of the Networked Society, along with Personal Assistants, the Internet of Things and Ambient Intelligence, all aimed at relations with consumers. The commercial relationships will have to be rethought and reshaped, and Privacy and Data Protection will be at stake. It is the moment to look at the right to be forgotten from a perspective of the B2C commercial relationship.

11.1 Introduction

Electronic Data Interchange has been around for more than 20 years and has been mainly used in Business to Business relationships. In the absence of a clear legal framework for the establishment and management of EDI relationships, soft law assumes a relevant role, especially the European Model EDI Agreement arising out of the Recommendation 94/820 EC Commission Recommendation of 19 October 1994 relating to the legal aspects of electronic data interchange. As the EDI relationship has traditionally been established in B2B relations, the key points for EDI negotiation have been on the one hand, the Interchange Agreement (with a legal, technical and security specification) [8, 15] and, on the other hand, subsidiarily, the referred EDI Model Agreement. Negotiation will thus play a crucial role in the design of the contractual regulation of aspects such as general management rules,

F.P. de Andrade (✉) · T.C. Moreira
University of Minho Law School, Braga, Portugal
e-mail: fandrade@direito.uminho.pt

T.C. Moreira
e-mail: tmoreira@direito.uminho.pt

M. Bundin · A. Martynov
Lobatchevsky State University, Nizhni Novgorod, Russia
e-mail: mbundin@mail.ru

A. Martynov
e-mail: avm@unn.ru

A. Costa et al. (eds.), *Personal Assistants: Emerging Computational Technologies*, Intelligent Systems Reference Library 132, DOI 10.1007/978-3-319-62530-0_11

liabilities, issues of proof, and also data protection. However, the transition of EDI to a new level, with the growing adoption of EDI in consumer relationships (consumers EDI) will bring along the weakening of the position of the consumer, most often totally incapable of any negotiation on the moment of drafting the contracts that will regulate the relationship [7]. The fragility of the consumer's position will further be evidenced if we consider the arising of a totally networked society, based on use of Personal Assistants and on the new concepts of the Internet of Things and Ambient Intelligence. This new framework of the relationship between the service provider and the consumer requires a new analysis of the EDI relationship and, along with it, of the technical and legal aspects of Ambient Intelligence and Data Protection, particularly the right to be forgotten.

11.2 Electronic Data Interchange

EDI is one of the techniques for interchanging data between interconnected computers. EDI means Electronic Data Interchange, being understood as electronic transfer, between terminals, of data processed by computer, concerning a commercial or administrative transaction, using a commonly accepted standard to structure the data [18].

This transmission of data may occur with or without human intervention. However, one of the obvious advantages of EDI is the use of totally automatic mechanisms allowing an efficient reduction of costs, concerning both the time of transmission and the costs with human resources. In EDI data are previously structured according to a certain format or standard, thus having messages formats equivalent to the several documents used in commerce allowing computers from different brands and types, and even in different countries, to communicate [10].

In EDI communication, for each commercial operation there will be a certain format of message EDI. Thus, any computer may, immediately and without any human intervention, identify the content of such messages [24] and automatically act, according to the message that was received. In this way, in an EDI message, the informatics system may automatically generate an order of sell and/or buy [24].

The use of structured EDI messages brings along several advantages [27] for the users:

- total abolition of paper [23];
- complete automation of the cycle order, delivery, payment
- huge reduction of time spent with exchange of documents, as well as a significant reduction of the required time for archiving, searching, discovering documents;
- significant increase of reliability and security in the processes of data interchange, avoiding errors and improving the quality of the service;
- increase in productivity and competitiveness [21, 23];
- speed of data transfer [27].

Of course, the use of electronic data interchange has the further and valuable advantage of allowing the immediate performance of commercial operations, by sending messages between different departments of the corporation, between different corporations or even third parties interested or involved in the transaction [27]. Thus, it allows for faster and more reliable transactions, less likely to generate conflict or dispute.

Electronic Data Interchange has been in use for more than 20 years, mainly in legal relations B2B. However, the developments in the field of Artificial Intelligence bring along the promising perspective of EDI migration to B2C relationships. According to data published in CSO Computer World [6], in this year of 2017 there will be 8.400 million devices connected through the Internet of Things, being 63% of the services aimed for consumers. It is expected that by the year 2020 this number will increase to 20.400 million devices, including around 5.200 million devices of consumers (smartphones, portable and domestic devices).

11.3 Personal Assistants and Ambient Intelligence

EDI faces new possibilities with the introduction of Personal Assistants and Ambient Intelligence, bringing along an enormous increase in the use of digital data, from which it is possible to build knowledge. The growing use of networks, increasingly wireless, allows the connection and the obtaining of data from any person and object [20].

Ambient Intelligence is fostering the change. The use of intelligent tools and the perspective of the Virtual Residence is now real, and quite often described as a tool merely used for purposes of efficiency and well-being. Home systems are now able to perform different functions of intelligent analysis of the environment through the use of services for the collection, analysis and processing of data [22].

Domotics and Ambient Intelligence are closely related. And the impact in EDI for consumers may soon be breathtaking. But that will bring the real threat of an intensive and constant processing of personal data and of monitoring, with the corresponding possibility of constructing and maintaining personal profiles that can be derived from the data used (and interchanged) by the applications. The massive collection of data by the systems, the interconnection of more and more devices with the Internet of things, the possibility of making available and interchanging data between systems, devices and databases [17] allows an easy monitoring of the user's choices and activities.

Ambient Intelligence will thus be implemented in such a way that electronic entities (Personal Assistants) will become active subjects, and the human users face the risk of becoming electronic entities built upon the generalized use of devices, programs and sensors. And as the software will have the perception of the surrounding environment and of the data transferred, it will be allowed a constant acquisition of data and information even about the contexts of interaction. The intensive use of more and more sophisticated technologies brings along an exponential increase in the possibilities of monitoring, surveillance and data control, with the consequent risk of dataveillance [17]. The risk is not limited to the access to data, but also to gathering information, crossing data and information, building up and using user profiles with the actual threat of a permanent construction of knowledge (and even of knowledge related to context).

11.4 Privacy and Data Protection

The transition of Electronic Data Interchange to Business to Consumer relationship, together with the implementation of Personal Assistants and Ambient Intelligence, brings along an exponential growth in the risks concerning privacy and data protection. Privacy is particularly threatened. The collection of data, the data mining and the building of user profiles, the sensors and the increased possibilities of monitoring, while they may be quite interesting for marketing purposes, brings along a progressive aggression in relation to personal rights and the consequent shadowing of the distinction between the public and the private sphere, thus enhancing the danger of Dataveillance [17].

Among the rights of the citizens in the electronic relationships we should note the right to be forgotten and to be let alone [12]: data must be preserved only while it is needed according to the purposes of collection and processing. There must be an adequate period of time for the preservation of data, in order to avoid a perpetual appropriation of very broad aspects of the personal life of the data holder [13]. And this must lead us to the consideration of a right to informational self-determination [12].

Yet, one thing is the doctrinal development of the right to be forgotten, another thing will be its practical implementation. Actually, in the Internet there is no right to be forgotten. Once the information is placed online it may last in cyberspace for a very long time, facing the risk of becoming totally out of date and inherently out of context. In theoretical terms, this right to be forgotten is linked to an evident problem of the digital age, since it becomes very hard to escape from the past in the Internet: every photo, every state update or tweet, or every commercial command may last forever in the "cloud" [4].

The right to be forgotten is being broadly debated and not only in Europe. Its origin, nature and content are still being interpreted in different ways. The most important issue here is the role of the right to be forgotten for privacy and privacy rights. Sometimes there is certain misunderstanding in comparing it with some already existing theoretical concepts and privacy rights.

Some scholars and practitioners have different views with regard to the nature of the right to oblivion. Most of them treat it as an European (French) concept droit à l'oublie that represents generally limits for media activities, forbidding press and TV to make public aspects of personal life mostly with a huge negative connotation that were the object of public interest in the past.

Those rights mentioned in Article 12 (right to erasure and blocking) and Article 14 (right to object) of Directive 95/46/EC5 are closely connected with data protection. They enable an individual to stop unlawful use of his data by a data controller. Those rights are closely connected with the idea that the data subject has the right to control data processing including the right to demand their blocking, erasure or to object to their processing by a controller [9, 11].

The ECJ, in its Decision Google Spain versus AEPD [16], has somewhat enlarged the scope of Directive 95/46/EC and applied it to search engines treating them as data controllers. There is no need to describe here all the details of the proceedings or their influence on EU practices, but it should be underlined for a better understanding of the state of things, that the Court clearly stated that search engines, even non-EU companies with sales and marketing subsidiaries inside the EU, are subject to the European Data Protection law. Obviously, the main aspect to be stressed is the right of the individual to control his data processing by third parties and treating personal data as part of his identity.

Considering the clear and present dangers citizens are facing, the European Commission started in 2012 a "package of measures" that were aimed at modifying the regulation of Data Protection in Europe, presenting a Proposal of Regulation on the issue. This Proposal was intended to revoke Directive 95/46/CE, currently still in force, but that will finally be revoked in May 2018, due to the approval, meanwhile, of the General Data Protection Regulation (GDPR) [5].

Directive 95/46/CE expressly establishes in Article 17 the right to be forgotten and the right to erasure, specially intended to deal with the issues related to the existence of online profiles and the several problems related to the option to quit an online profile.

The new Regulation also intends to affirm and enhance a right that already exists (but actually has no practical efficacy), allowing that someone's profile ceases to exist and not merely stays in a sort of hibernation. Obviously, this does not mean that it supports a right to the total elimination of history. What is intended to be deleted is the personal information that someone gave about his/her own person, but not the references about him/her that appeared in the media, for information purposes or whenever there are legitimate motives for its conservation.

This right to be forgotten is intended to be a right of defense of every person, being assumed as a right of control of his/her personal data, allowing to control its online availability, regardless of whether its divulgation was, or was not, authorized. This right intends that corporations responsible for social networks and other information society services (including electronic commerce) erase all personal data of the users when these cancel the service. This right must include the removal of all data in the Internet pages where these are included and the elimination of any reference to the said data in search engines.

On the other hand, the aim is that users should have access to their personal data and that they should be able to easily transfer them, as well as impose limits concerning the period during which service providers may keep information on the users, the quantity of data visible online after their removal has been requested and the right to sue the site owners in case of non-compliance with the user's request.

Surely, the right to be forgotten may represent a strong and effective legal mechanism for protecting individuals privacy in digital era. Its core element as it has been already said here is in recognition of an individuals freedom to control information about him. It is beyond any doubt that a person should have a right to start living with a clean slate, especially when the criminal record is expired or he was acquitted. For example, EU Commissioner Viviane Reding described the ECJ in Google Spain versus AEPD as "a clear victory for the protection of personal data of Europeans" [25]. This point of view received a lot of support throughout Europe and its rationale is clearly understandable.

Nevertheless, implementation of such a right will affect freedom of information and transparency that are one of the most important elements of information society. Vast majority of requests to Google and probably to other searching engines stresses the issue about the consequences for information society and transparency. Moreover, a very appropriate question about probable misuse of the right to be forgotten to hide socially important information from the public should be always kept in mind.

This issue was strongly debated in Great Britain and was named by some MPs as "a draconian attack on free speech and transparency, totally at odds with Britains liberal tradition" [14]. Open Rights Group also expressed significant concern about the consequences to online free speech that may follow the ruling [1].

Further implementation of the right to be forgotten practice by Google also evoked a great concern from a large number of scholars and practitioners in Europe. In May 2015, a large group of 80 academics published an open letter [19] to Google asking to make its delisting policies and decision-making processes more transparent to the public and data protection authorities. They name the existing mechanism a silent and opaque process, with very little public process or understanding of delisting.

Within a day or two of the ruling, already the stories were coming out about paedophiles, politicians and general practitioners wanting to use the right to be forgotten to delete their Internet stories a sort of rewriting of history that the term right to be forgotten evokes, but that the ECJ ruling in its nature does not provide for [26]. In this case, a vast majority of academics and politicians argued against such a use of the right arguing that it do not concern public figures or information of great social value.

All this statements are true but whether it is for Google to decide if the information is irrelevant, inadequate or no longer relevant? That question could not be easily answered. In their letter, 80 academics in fact admitted not only the necessity of transparency in delisting information but also the necessity of a public or data protection authoritys control other the process of delisting in order to find an appropriate balance between individual privacy and public discourse interests.

Although the consecration of this right to be forgotten must be considered as clearly positive, one should also take into account the criticism on the part of the Group of Article 29 on this subject [2]. Thus, the person responsible for the processing of data is liable for the erasure of the data but also for the communication to a third party that eventually will process the data through links (hyperconnections), copies, reproductions (not forgetting that service providers of information society are considered responsible for the processing of data, on online social networks, since they provide the means for the processing of the users data and also provide all basic services related to users management such as registration and cancellation of accounts [3]. Thus, to impose this obligation only on the person responsible for the processing may bring along some problems. It may well happen that this person has adopted all reasonable measures to inform third parties but, meanwhile, this person may have no knowledge of all the existing copies or reproductions, or even the possibility of new copies or reproductions arising after the person responsible for the data has informed all third parties.

Furthermore, as it is well noticed by the Group of Article 29, there is no disposition at all making it mandatory for the third parties to comply with the request of the data holder, unless they should also be considered responsible for the processing of the data.

Furthermore, the Regulation does not provide any guidelines concerning the way how the holders of personal data may exercise their rights in case of the disappearance of the person responsible for the processing (or if he can no longer be contacted). In this sense, it must be said that the legal position of third parties processing personal data must be clarified, by extending the rights of the data holders so that they may send the request of erasure directly to third parties (in case it is not possible to do it through the person responsible for the processing of personal data) [2].

We must agree with this legal consecration of this right, since the providers of social networks have to comply with legal duties that make effective the power of control of their users and the free disposal of their personal data, regardless of the type. That means, regardless of being (or not) intimate data, in the broad sense of personal data considered both in the current Directive of Data Protection or in the GDPR.

The right to privacy of the user of social networks and other information society services, while holder of the personal rights, as part of his (her) right to informational self determination, has constitutional grounds in Portugal, from Article 35 of the Portuguese Constitution, imposing on the service providers (that have, keep, use and process, in paper or in digital form, automatically or not, users personal data) the obligation to fulfil the requirements of the data protection legislation. Thus, the right to privacy as part of the right to informational self determination becomes the basis of the freedom and dignity of persons.

Undoubtedly, the right to be forgotten could represent a strong and effective legal mechanism for protecting individuals privacy in the digital era. Its core element as it has been already said here is the recognition of an individuals freedom to control information about himself.

11.5 Final Remarks

EDI has been around in commercial activities for more than 20 years. It is now expected its growth in business to consumers relationships, alongside with developments in the field of Artificial Intelligence. Personal Assistants, Internet of Things and Ambient Intelligence will gradually invade our home places and we will be surrounded by all sort of connected devices. And although technology will allow easier, faster and more reliable transactions, new dangers will arise out of the introduction of Personal Assistants and Ambient Intelligence. We will now live under the threat of an intensive and constant processing of personal data and of monitoring, with the corresponding possibility of constructing and maintaining personal profiles. Privacy is particularly threatened. Personal rights must be reconsidered and, among them, the right to be forgotten and to be let alone—data must be preserved only while it is needed according to the purposes of collection and processing. Directive 95/46/EC5 already enabled an individual to stop unlawful use of his data by a data controller. The ECJ, in its Decision Google Spain versus AEPD has somewhat enlarged the scope of Directive 95/46/EC and applied it to search engines treating them as data controllers. The new Regulation also intends to affirm and enhance the right of allowing that someone's profile ceases to exist. Regardless of some justified criticism on the way it is formulated, this right to be forgotten is intended to be a right of defense of every person, being assumed as a right of control of his/her personal data. A particular focus will now be on providers of social networks that shall have to comply with legal duties in order to make effective the power of control of their users and the free disposal of their personal data.

References

1. ECJ google spain ruling raises concerns for online free speech. https://www.openrightsgroup.org/press/releases/ecj-google-spain-ruling-raises-concerns-for-online-free-speech
2. Group of article 29, opinion 01/2012 on the proposals of reformation concerning data protection, adopted on the 23rd march 2012. http://ec.europa.eu/justice/data-protection/index_pt.htm
3. Group of article 29, opinion 5/2009, 12th June 2009, pp. 5–6
4. Group of article 29 opinion 05/2012 on cloud computing. http://ec.europa.eu/justice/data-protection/index_pt.htm
5. Regulation (EU) 2016/679 of the european parliament and of the council of 27 April 2016. http://eur-lex.europa.eu/legal-content/EN/TXT/HTML/?uri=CELEX:32016R0679Żfrom=EN
6. Cso-computer world (22 March 2017). http://cso.computerworld.es/archive/la-falta-de-estandares-supone-una-amenaza-para-iot
7. Almeida, C.F.: Direito do Consumo. Almedina (2005)
8. Andrade, F.C.P.: A celebração de contratos por EDI—Intercâmbio Electrónico de Dados. In: Estudos em Comemoração do 10 aniversário da Licenciatura em Direito da Universidade do Minho, pp. 297–322. Almedina, Coimbra (2004) (in Portuguese)
9. Ausloos, J.: The 'right to be forgotten'—worth remembering? Comput. Law Secur. Rev. **28**(2), 143–152 (2012). doi:10.1016/j.clsr.2012.01.006
10. Baum, M.S., Henry, H., Perritt, J.: Electronic Contracting, Publishing and EDI Law. Wiley Law Publications (1991)
11. Bundin, M., Martynov, A.: Legal perspective to be forgotten in russia and freedom of information. In: Communications in Computer and Information Science, pp. 169–179. Springer International Publishing (2016). doi:10.1007/978-3-319-49700-6_17
12. Castro, C.S.: Direito da Informtica: Privacidade e Dados Pessoais. Almedina (2005)

13. de la Cueva, M., Lucas, P.: Informtica e proteccin de datos personales. Centro de Estudios Constitucionales (1993)
14. Doughty, S.: Europe grants the 'right to be forgotten online': EU court will force google to remove people's personal data from search results on request. The Daily Mail (13 May 2014). http://www.dailymail.co.uk/sciencetech/article-2626998/The-right-forgotten-EU-court-rules-Google-remove-personal-data-search-results-request.html
15. Festas, D.D.O.: A contrata electrnica automatizada. In: Direito da Sociedade da Informa, vol. VI, pp. 411–460. Coimbra Editora (2006) (in Portuguese)
16. Floridy, L.: The right to be forgotten the road ahead. The Guardian (2014). https://www.theguardian.com/technology/2014/oct/08/the-right-to-be-forgotten-the-road-ahead
17. Hert, P.D., Gutwirth, S., Moscibroda, A., Wright, D., Fuster, G.G.: Legal safeguards for privacy and data protection in ambient intelligence. Pers. Ubiquit. Comput. **13**(6), 435–444 (2008). doi:10.1007/s00779-008-0211-6
18. Jones, P., Marsh, D., Walker, R.: Essentials of EDI Law. Blackwell Publishers (1994)
19. Kiss, J.: Dear google: open letter from 80 academics on 'right to be forgotten'. The Guardian (14 May 2015). http://www.theguardian.com/technology/2015/may/14/deargoogleopenletterfrom80academicsonrighttobeforgotten
20. Mills, M.P., Ottino, J.M.: The wall street journal (30 January 2017). https://www.wsj.com/articles/SB10001424052970203471004577140413041646048
21. Muro, J.D., Ramn, J.L.M., Rodriguez, J.S.: Servios bsicos de seguridad en la contractacin electrnica. In: Encuentros sobre Informtica y Derecho 1994–1995. Aranzadi Editorial (1995)
22. Oh, J.S., Park, J.S., Kwon, J.R.: A study on autonomic decision method for smart gas environments in korea. In: Advances in Intelligent and Soft Computing, pp. 1–9. Springer Berlin Heidelberg (2010). doi:10.1007/978-3-642-13268-1_1
23. Reed, C.: Computer Law. Blackstone Press Limited (1990)
24. Reed, C.: Advising clients on EDI contracts. In: Computer Law and Practice, vol. 10 (1994)
25. Timberk, C., Birnbaum, M.: In google case, EU court says people are entitled to control their own online histories. Washington Post (13 May 2014). https://www.washingtonpost.com/business/technology/eu-court-people-entitled-to-control-own-online-histories/2014/05/13/8e4495d6-dabf-11e3-8009-71de85b9c527_story.html
26. Wakefield, J.: Politician and paedophile ask google to 'be forgotten'. http://www.bbc.com/news/technology-27423527
27. Walden, I.: EDI: contracting for legal security. In: International Yearbook of Law, Computers and Technology, vol. 6 (1992)

Chapter 12
Personal Assistants: Civil Liability and Dispute Resolution

Marco Carvalho Gonçalves

Abstract In this chapter, we will study the problem of civil liability resultant from the use of intelligent personal assistants and the most appropriate legal means to resolve disputes in this area. Thus, starting from a preliminary analysis with regard to the protection of personal rights, we will analyse the different illicit behaviours that may occur as a result of the use of intelligent personal assistants, as well as the legal requirements for the liability that have to be verified and the possible need to implement new legislation on this area. In the second part of our study, we will examine the different legal means that can be used to resolve civil liability disputes arising from the use of intelligent personal assistants, with a particular focus on alternative means of dispute resolution.

12.1 Introduction

In the past years, there has been a gradual aging of the world population due to the increase in the average life expectancy and the decline in the number of births.

This generalized aging of the population places special challenges to modern societies, particularly due to the need to guarantee people's quality of life, without neglecting the sustainability of the social systems.

Besides, in addition to the problem of the general aging of the population, another challenge that calls for an urgent and adequate response is the need to ensure better medical and social care for people suffering from diseases that are susceptible to decline their physical and psychological abilities and compromise "the completion of activities of daily living" [1].

However, for reasons of a different nature, particularly in the economic and social domain, it may occur that such aid is not possible or, at least, not provided as effectively as it would be desirable.

M.C. Gonçalves (✉)
School of Law, Universidade do Minho, Braga, Portugal
e-mail: marcofcg@direito.uminho.pt

A. Costa et al. (eds.), *Personal Assistants: Emerging Computational Technologies*,
Intelligent Systems Reference Library 132, DOI 10.1007/978-3-319-62530-0_12

In this regard, the concentration of economic activities in urban centers and the intensification of migratory movements has been increasing the isolation of populations, so that many older people live alone and away from their relatives [2].

Furthermore, older or debilitated people may live in very isolated places, where there are no health services or institutions that can provide care.

In any case, it is curious to observe that even in situations where older or debilitated people live in isolated or distant places from their more direct relatives, they still prefer to continue to live in their homes than in specialized institutions in personal care.

This reality poses to the modern societies a challenge that needs to be answered urgently, consubstantiated in the need to guarantee the elder or the debilitated a dignified, healthy and safe life.

The initial response to this challenge was the development or the implementation of institutions specialized in providing health care or social solidarity.

However, this response turned out to be unsatisfactory, taking into account the quick aging of the population and the subsequent increase in the demand for such services at levels far above from those offered or because of the very high costs associated with the provision of health and social solidarity services.

It was precisely in this context that, due to the bankruptcy of the solution initially adopted in relation to the growing needs in the provision of health care and social solidarity, it has been implemented in recent years the use of information and communication technologies [3]—like "communication devices; sensory aids; consumer electronics/multimedia; smart home technology; medical assistive technology; tele-monitoring devices; walking aids and assistive devices for supporting mobility; automatic and intelligent devices and services capable of helping older individuals performing their daily chores" [4] —, particularly, intelligent personal assistants [5], seeking to improve the quality of life and the well-being of older or disabled people [6].

Indeed, intelligent personal assistants are able to guarantee, with minimal costs, an active, safe and healthy aging at home [7], avoiding the need of institutionalization [8].

On the other hand, even in relation to people who, due to their advanced age or physical or psychological weaknesses, present greater difficulties of locomotion and consequent tendency to live in greater isolation, the implementation of intelligent personal assistants is likely to guarantee greater autonomy, independence, quality of life and safety, by "monitoring and assessing care needs" [9].

Indeed, the use of intelligent multi-agent systems contributes to greater autonomy and independence of people, and allows the reducing of health costs [10], which is particularly relevant in the context of a society with a population growing older [11].

In this regard, intelligent personal assistants can perform many relevant tasks, among which are the following: warn about the need to take a particular medication and the respective dosage [12]; advise or assist the user to do shopping, buy goods and receive them directly at home, without the need for any physical movement [13]; find the house keys; regulate the illumination or temperature of a space;

automatically detect if the cooker is switched on inadvertently; monitor accidental falls [14], perform biometric tests [15]; make automatic telephone calls to a family member or an emergency service in the event that the person needs urgent medical intervention, etc.

In any case, the use of intelligent personal assistants may raise some legal problems, particularly in the field of protection of personality rights and in matters of civil liability. For that reason, it is also necessary to examine the most appropriate judicial or extrajudicial means of resolving disputes that may arise in this area.

12.2 Protection of Personality Rights

The great generality of modern legal systems contains legal norms to protect people against any offense to their physical or moral integrity. Under the Portuguese legal system, the general protection of personality is provided by the articles 24 to 26 of the Constitution of the Portuguese Republic, which establish the inviolability of human life and the protection of physical and moral integrity, as well as the constitutional protection of the rights to personal identity, personality development, civil capacity, citizenship, good name and reputation, speech, privacy of private and family life and legal protection against all forms of discrimination. Likewise, the Constitution of the Portuguese Republic provides that the law must establish effective guarantees against the obtaining or abusive use or misconduct of human or relative information concerning a person or a family.

These constitutional principles find expression in the article 70 of the Portuguese Civil Code, which states that "the law protects individuals against any unlawful or moral offense against their physical or moral personality".

Thus, in the event of a violation of the physical or moral personality, the offended person, in addition to the civil or criminal liability that may arise, may also request the adoption of appropriate measures to the circumstances of the case, in order to avoid the consummation of the threatening or to mitigate the effects of that offense.

In this way, the protection of personality rights can be guaranteed through the adoption of precautionary measures, through the regime of criminal responsibility or through civil liability of the offending agent. In this chapter, we will deal only with the matter of civil liability.

12.3 Civil Liability

12.3.1 Introduction

Despite their importance and their recognized capabilities and potential, the use of intelligent personal assistants can lead to civil liability disputes.

Indeed, in the particular case of the use of intelligent personal assistants in the field of information and communication technologies, civil liability may result either from the infringement of subjective rights, in particular personality rights, namely the right to life, physical or moral integrity and the reservation of physical or moral integrity, or from the non-fulfilment of obligations associated with the specifically contracted service.

On the other hand, such civil liability may result from an action or an omission.

Generally, the violation of rights will result from the adoption of a particular wrongful conduct. This is the case, for example, with the unauthorized gathering, processing or disclosure of personal data, as well as the transmission of false or misleading information or recommendations (e.g. wrong information regarding the medication to be taken or in relation to its dosage), with the consequent production of damages.

It may, however, occur that this breach is also a result of the omission of a legal or contractual conduct [16]. Accordingly, civil liability for omission may be incurred in respect of damage caused by failure to comply with the obligation of supervision or assistance, in particular as a result of failure to detect situations of risk or potential danger to the person.

In addition, in the field of intelligent personal assistants, the occurrence of damages may be particularly serious because of the reliance placed on them by the users of such services or by their family members, who would be subject to supervision or assistance. In fact, as with institutional services for medical care and social solidarity, the use of intelligent personal care services is likely to give the user and his family a special confidence in the proper functioning of these services, with the consequent reduction of the critical spirit regarding the actions, recommendations or suggestions presented by personal assistants [17].

In this sense, only through the protection of this trust can be guaranteed that these technological services may be seen as a safe alternative to the institutionalization of the elder or the sick. In fact, according to the principles of appearance and trust, "those who create by their activity and behaviour in legal relations, expectations of reliability and security, must bear the consequences of the frustration of these values" [18].

12.3.2 Types of Illegal Conduct

12.3.2.1 Unauthorized Use and Processing of Personal Data

In accordance with the articles 12.º of the Universal Declaration of Human Rights, 8.º of the European Convention on Human Rights, and 7.º and 8.º of the Charter of Fundamental Rights of the European Union, no one shall be subjected to arbitrary interference with their privacy, their family, at home or in their correspondence [19]. This fundamental right to "self-determination" guarantees the protection of the person against the State, which can intervene only through "legal forecasts and authorizations, while also respecting the principle of proportionality" [20].

Thus, the possibility of using intelligent personal assistants presupposes that the user gives its informed consent to the access, treatment and possible disclosure or free circulation of its personal data. This is the case, in particular, with information regarding to its name, age, gender, clinical or biometrical data, location and personal contacts, as well as the capture or reproduction of photographs or video [21].

In fact, everyone has the right to self-determination information, that is to say, to have control over his personal data and the protection of his privacy [22], so that a violation of human rights can result in civil or criminal liability.

Therefore, the use of intelligent personal assistants necessarily depends on the prior and informed consent of the user or, in the case of incapacity, on the person who exercises legal powers of representation, under penalty of civil liability for violation of personality rights [23].

12.3.2.2 Violation of Life or Physical Integrity

According to the article 3.º of the Universal Declaration of Human Rights, everyone has the right to life, liberty and security of person. The same protection arises from the article 2.º of the European Convention on Human Rights, according to which the right of every person to life is protected by law.

Indeed, the right to life is the principal right of the human being and should, therefore, be protected by the State, since it is an inviolable right [24].

In any case, the right to life "is not an absolute right, and it must be understood that limitations must be interpreted strictly in the light of the principle of the inviolability of human dignity" [25].

Intelligent personal assistants can play a key role in protecting life and human dignity. As a matter of fact, the use of these information and communication technologies is, as mentioned, capable of guaranteeing greater security and quality of life for people. It is enough to think on the possibility of intelligent personal assistants allow to detect the need for medical care, control the health parameters of users or assist in the provision of basic hygiene and health care.

However, the use of intelligent personal assistants can also cause offense of life or physical integrity. It is enough to think of situations in which, by default, personal assistants do not adequately perform the functions for which they are programmed, resulting of this omission the offense of physical or moral integrity or the death of the person to be assisted.

12.3.3 Appreciation of Fault and Damages

With regard to fault, civil liability attributable to the entities responsible for the management or control of intelligent personal assistants may be due in deceit or neglect.

In effect, in the first case, the agent deliberately acts in view of the production of an injury, that is to say, this modality of guilt presupposes the existence of an intellectual element, consubstantiated in the representation of a certain fact that the agent knows to be illicit, as well as of a voluntary element, translated in the effective will of the agent in the sense of acting with the intention to commit the wrongful act, accept the production of the wrongful act as a necessary or normal consequence of his conduct or conform to the actual production of the damage as a consequence of his unlawful performance.

In turn, in the second case, civil liability will arise from the omission of a general duty of conduct, that is, the agent acts without observing the duty of care which, under normal conditions, he should have ensured. Thus, negligence may occur or because the agent, even though he represented a particular wrongful act as possible, has confided that such fact would not occur or because he did not even represent as possible the production of an unlawful act, even though he should have a duty to take account of this possibility.

In the particular case of the use of intelligent personal assistants, the typical fault will be negligent [26], since it is difficult to think of the possibility of an intelligent personal assistant acting in an intentional way, that is, with the concrete purpose of producing an illicit and harmful event. Fall within this scope the situations in which a failure in monitoring or surveillance occurs, leaving a person defenseless or exposed to dangerous situations, as well as situations in which, in violation of the general duties of care, the collection, use or improper treatment of personal data occurs. In any case, the punishability of negligent performance will necessarily fall not on the intelligent personal assistant, but rather on the entity responsible for the development, design or application of the intelligent personal assistant.

Relativity to damages, in general terms, two types of damages can occur: property damages or moral damages.

In the first case, the damages can be quantified in economic terms, and these damages can be classified as damages that arise directly, that is, damages that result directly from the unlawful and culpable action, or as loss of profits, that is, the losses suffered by the injured as a result of this illicit and guilty act.

In turn, moral damages are those that, not properly affecting the heritage of the injured party, are not susceptible of being valued in equity, because they respect values of a moral or spiritual nature. This is the case, for example, with a breach of the right to a name or image. For this reason, the Portuguese legal system determines that such damages must be repaired through the use of equitable criteria.

The use of intelligent personal assistants is liable to cause damages of different nature, whose repair will depend, in any case, on the existence of a causal link between the illicit and guilty conduct and the production of damages. Indeed, it is possible to think, by way of example, of the damage caused by unlawful or improper disclosure of personal data, wrong reading of biometric data and subsequent violation of the right to life or to physical integrity [27], breach of the duty of supervision or care, omission of duty to provide assistance or the provision of incorrect or false information.

In any case, the progressive implementation in our societies of intelligent personal assistants poses difficult problems with regard to the determination of possible civil liability, since technological systems will naturally be designed to protect the personality and physical and moral integrity.

Thus, it appears that the traditional regime of civil liability, which implementation depends on the verification of very specific substantive requirements, resulting in the occurrence of an unlawful, culpable and harmful fact, as well as in the verification of a causal link between the illicit and guilty action and the damage, may not be adequate to this new technological reality.

12.4 Dispute Resolution

The progressive use of intelligent personal assistants is likely to lead to the emergence of litigation, particularly in the areas of civil liability and consumption. In this sense, it is necessary to adapt both the legislation and the means of resolving disputes to this new reality.

First of all, one of the main problems concerns the possible lack or insufficiency of legislation on intelligent personal assistance. Indeed, the rapid development of this communication and information technology requires the adoption of specific legislation that adequately regulates the use and the introduction of the intelligent personal assistants in the market, the entities responsible for their management and the rights and obligations associated with the provision of such services, particularly consumer protection. In fact, the absence of any specific legislation in this area will entail the verification of legal gaps or the application of general legal institutes, which may be inadequate or inappropriate to these new technological realities.

On the other hand, with regard to the dispute resolution, it is important to consider the problem of legitimacy, that is to say, the identification of the entity that may be sued in judicial or extrajudicial proceedings.

In fact, liability for any breach of statutory consumer protection laws or for the verification of unlawful actions or omissions, sanctionable under the civil liability law regime, will fall directly on the entity responsible for the creation, development, implementation or management of the intelligent personal assistant.

However, such legitimacy may also extend to other entities, such as importers, national representatives of manufacturers, agents, traders or distributors.

Moreover, it may also happen that this passive legitimacy rests on several entities simultaneously. It is enough to think, for example, on the possibility that an intelligent application may require other accessory instruments (e.g. image capture, temperature or motion sensors, etc.), so that the production of the damage results of a joint or simultaneous failure of several elements or components.

With regard to the active legitimacy, that is to say, the identification of the person who may be demanding in a judicial or extrajudicial procedure, that legitimacy belongs to the consumer or the injured party, depending on how the judicial or extrajudicial procedure is configured or qualified.

On the other hand, the resolution of disputes related to the use of personal assistants raises important issues concerning the identification of the most appropriate legal procedures to solve those disputes.

In fact, the traditional way of settling disputes in the field of civil liability or in the domain of consumer relations is the recourse to judicial courts.

However, the use of judicial remedies raises some problems which need to be emphasized.

First of all, one of the main problems concerns the determination of the court having jurisdiction to settle the dispute. This is because disputes relating to the use of systems integrated in information and communication technologies will normally be associated with international legal relationships, that is, which are in contact with different legal systems of different States. This is the case, for example, if the responsible for the development of an intelligent personal assistant is domiciled in a particular State and the consumer or the injured party is domiciled in another State.

In this context, under European law, the Regulation (EU) n.º 1215/2012 of the European Parliament and of the Council of 12 December 2012 regulates the jurisdiction, recognition and enforcement of judgments in civil and commercial matters.

Thus, in the case of a consumer dispute (for example, the product or service purchased or contracted does not have the technical qualities that were presented to the consumer), shall have jurisdiction the court of the State where the defendant has his residence or the court of the place of the consumer's residence, regardless the domicile of the other party.

In turn, in a civil liability case, the court of the defendant's residence or the court of the place of the harmful event, that is to say, of the place where the damage occurred, is competent for the judgement of the dispute.

In any case, the physical distance between the parties or between the parties and the court, as well as the lack of knowledge of the legal regimes applicable to the resolution of the particular dispute, may be a disincentive for the use of traditional judicial remedies.

Another aspect that should be emphasized is the technical difficulty associated with the resolution of these disputes by the traditional judicial process.

In fact, since these conflicts are associated with the application or development of information or communication technologies, it may occur that the judge does not have the necessary qualifications or the technical knowledge to judge the dispute, which may imply that he has to seek assistance of specialized technical professionals to allow him to verify if, in the specific case, there is a violation of legal norms or the practice of any infraction that could justify civil liability.

Furthermore, the judgment of disputes relating to intelligent personal assistants may also bring special difficulties in evidential matters, requiring the intervention of specialized professionals, resulting in unwanted delays to the judicial process. This will be the case, in particular, of the need for inspections or judicial expertise, without neglecting that it may not be possible to carry out such proceedings because the physical distance between the parties in relation to the court having jurisdiction to hear the case.

The difficulties associated with the resolution of disputes through judicial courts may be overcome through the use of alternative means of dispute resolution, namely conciliation, mediation, negotiation and arbitration.

In fact, these alternative means of dispute resolution, which have been progressively adopted in the different Member States of the European Union, have the advantage of being simpler, easier and quicker means compared to judicial proceedings. These factors are very important when choosing the most appropriate means of resolving a particular dispute, since the effective remedy of a right that has been violated depends substantially on the speed of such reparation as a manifestation of the principle of effective protection.

In particular, mediation has great potential as a means of out-of-court settlement of disputes concerning the use of intelligent personal assistants, in that, through recourse to a mediator, as an impartial, independent and indifferent entity in relation to the dispute, the parties shall, by means of a simple, confidential and voluntary procedure, communicate with each other intending to reach an agreement enabling the resolution of the dispute.

Moreover, in the particular field of consumer relations, mediation has played a key role, in particular by allowing the resolution of disputes on the basis of fair and friendly solutions.

However, despite of its importance, mediation developed in a physical, personal and face-to-face environment may not be adequate to the resolution of disputes related to the use of intelligent personal assistance technologies.

In this context, online dispute resolution has several potentialities that traditional models of judicial or extrajudicial conflict resolution can't guarantee. This is the case, for example, with the possibility of resolving disputes without the need for physical movement of the parties, using deferred communication, like videoconference, teleconference, online or offline sessions [28], as well as the implementation of simple and easy-to-understand rules allowing litigants to, quickly and effectively, find a solution to their dispute [29].

These advantages are particularly important in low-cost litigation [30].

Moreover, since the litigation concerns directly or indirectly the use of information technology, it is a particularly important factor that this litigation can also be resolved through the use of technological systems [31], which guarantee a simple, flexible, fast and confidential procedure.

On the other hand, the use of online means of settling disputes—for example, the online dispute resolution platform for consumer disputes [32]—allows the quick and economical resolution of disputes, which is particularly important in cross-border disputes.

Lastly, it can't be ignored that online dispute resolution is better adjusted to the very use of personal assistants integrated in communication and information technologies, as it allows a faster, simpler and more effective presentation of the complaint and facilitates the dialogue between the parties, allowing the resolution of the dispute within a reasonable time and in accordance with equitable solutions.

12.5 Conclusion

In recent years, there has been a gradual aging of the population, which poses special challenges to modern societies, given the need to guarantee the quality of life of people and the sustainability of the social systems.

In this sense, intelligent personal assistants are an alternative to the institutionalization of people, besides providing them with a higher quality of life and well-being.

However, in spite of the importance and the potential of intelligent personal assistants, the truth is that the use of this technology can lead to civil liability conflicts, which can result either from the violation of subjective rights—e.g. unauthorized processing of personal data or violation of life or offense to physical integrity—or failure to comply with obligations associated with the service actually contracted.

On the other hand, the rapid development and implementation of this communication and information technology requires the adoption of specific legislation.

In addition, the resolution of disputes relating to the use of intelligent personal assistants through the judicial courts may raise specific problems, in particular as regards the determination of competence and legitimacy, as well as technical difficulties for which the court may not to be prepared.

Therefore, the difficulties associated with the resolution of disputes through judicial courts may be overcome through the use of alternative means of dispute resolution, namely conciliation, mediation, negotiation and arbitration, which, in addition to being simpler, faster and effective procedures, also have the advantage of allowing online dispute resolution, which is the most appropriate procedure in the field of information and communication technologies.

References

1. Mihailidis, A., Carmichael, B., Boger, J.: The use of computer vision in an intelligent environment to support aging-in-place, safety, and independence in the home. http://ieeexplore.ieee.org/abstract/document/1331400/metrics. Accessed 30 Mar 2017
2. Torta, E. et al.: Attitudes towards socially assistive robots in intelligent homes: results from laboratory studies and field trials. J. Hum.-Robot Interact. **1**(2), 77 (2012)
3. Cesta, A. et al.: Supporting Interaction in the ROBOCARE Intelligent Assistive Environment. https://vvvvw.aaai.org/Papers/Symposia/Spring/2007/SS-07-04/SS07-04-004.pdf. Accessed 29 Mar 2017
4. Nani, M. et al.: MOBISERV: An Integrated Intelligent Home Environment for the Provision of Health, Nutrition and Mobility Services to the Elderly. http://eprints.uwe.ac.uk/16102/1/MOBISERV_WorkshopPaper.pdf. Accessed 20 Mar 2017
5. This is the case, for example, with the personal assistants "Living well with Anne", "Home4Dem", "Care4Balance", "Confidence", "Hearo" and "iSense". There are also several applications for smartphones, such as "Daisy Care" and "Lumen Health"

6. Final Evaluation of the Ambient Assisted Living Joint Programme, European Union, p. 4 (2013). http://www.aal-europe.eu/wp-content/uploads/2015/05/Final-report-of-the-AALBusquin-2013.pdf. Accessed 28 Dec 2016
7. Decision n.º 554/2014/EU of the European Parliament and of the Council of 15 May 2014, on the participation of the Union in the Active and Assisted Living Research and Development Programme jointly undertaken by several Member States
8. Pieper, M., Antona, M., Cortés, U.: mbient Assisted Living. in Ercim News **87**, 19 (2011)
9. Blanson Henkemans, O.A., et al.: Usability of an adaptive computer assistant that improves self-care and health literacy of older adults. Methods Inf. Med. (2008)
10. Costa, R. et al.: Ambient Assisted Living. http://www4.di.uminho.pt/~pjn/Projectos/FCT2009/Pub/UCAMI2008.pdf. Accessed 27 Dec 2016
11. See, in this regard, the Communication from the European Council of October 12, 2016, entitled "The demographic future of Europe—from challenge to opportunity"
12. van den Broek, G., Cavallo, F., Wehrmann, C. (ed.), AALIANCE Ambient Assisted Living Roadmap, pp. 10–11. IOS Press (2010)
13. Gilbert, D.: Intelligent agents: the right information at the right time, p. 3. https://fmfi-uk.hq.sk/Informatika/Uvod%20Do%20Umelej%20Inteligencie/clanky/ibm-iagt.pdf. Accessed 30 Mar 2017
14. Laguna, M.A., Finat, R.: Remote monitoring and fall detection: multiplatform java based mobile applications. In: Bravo, J., Hervás, R., Villarreal, V. (ed.) Ambient Assisted Living, pp. 1–2. Springer (2011)
15. SILVA, Bruno M. C./RODRIGUES, Joel J. P. C./LOPES, Ivo, M. C. de M., "Pervasive and Mobile Health Care Applications", in Ambient Assisted Living, Nuno M. Garcia, Joel J. P. C. Rodrigues (ed.), CRC Press, 2015, p. 42
16. FONSECA, Tiago Soares da, "Da tutela judicial civil dos direitos de personalidade", in Revista da Ordem dos Advogados. http://www.oa.pt/Conteudos/Artigos/detalhe_artigo.aspx?idsc=47773&ida=47781. Accessed 30 Dec 2016
17. Huyck, C., Augusto, J., Gao, X., Botía, J.A.: Advancing ambient assisted living with caution. In: ICT4 Ageing Well, M. Helfert et al. (eds.), p. 21. Springer (2015)
18. Judgment of the Portuguese Supreme Court of Justice, of April 10, 2014, proc. 8476/03.8TBCSC.L1.S1, in http://www.dgsi.pt/jstj.nsf/954f0ce6ad9dd8b980256b5f003fa814/7942ff76dd1b702980257cba003631b5?OpenDocument (accessed in 02.01.2017)
19. See the Directive 95/46/CE of the European Parliament and of the Council of 24 October 1995 on the protection of individuals with regard to the processing of personal data and on the free movement of such data
20. Ruaro, R.L., Rodriguez, D.P., Finger, B.: O direito à proteção de dados pessoais e à privacidade. In: Revista da Faculdade de Direito—UFPR, n.º 53, Curitiba, p. 56
21. Judgment of the Portuguese Supreme Court of Justice, of June 7, 2011, proc. 1581/07.3TVLSB.L1.S1, in http://www.dgsi.pt/jstj.nsf/954f0ce6ad9dd8b980256b5f003fa814/6190a514bc9d85cf802578aa003183d3?OpenDocument (accessed in 28.12.2016)
22. Judgment of the Portuguese Supreme Court of Justice, of October 10, 2014, proc. 679/05.7TAEVR.S1, in http://www.dgsi.pt/jstj.nsf/ 954f0ce6ad9dd8b980256b5.f003fa814/c3ca422fd42da4ba80257de0003469cf?OpenDocument (accessed in 28.12.2016)
23. Allen, J.E.: Assisted Living Administration: the knowledge base, 2nd. ed., Springer, New York, p. 272
24. ROBERTO, Luciana Mendes Pereira, Responsabilidade Civil do Profissional de Saúde & Consentimento Informado, 2nd. ed., Juruá Editora, Curitiba, 2008, pp. 34–35
25. Silveira, A., Canotilho, M. (ed.), Carta dos Direitos Fundamentais da União Europeia, Almedina, Coimbra, 2013
26. Doolan, S., Agosto, M.: Assisted living cases: investigation and pretrial considerations. In: Patricia, W.I. (ed.) Nursing Home Litigation: Investigation and Case Preparation, 2nd. ed., p. 57. Lawyers & Judges Publishing Company (2006)
27. O'Grady, M.J., Muldoon, C., Dragone, M., Tynan, R., O'Hare, G.M.P.: Towards evolutionary ambient assisted living systems. J. Am. Intell. Humaniz. Comput. Springer,

p. 3. http://irserver.ucd.ie/bitstream/handle/10197/1915/ogrady_jaihc_p22.pdf?sequence=1. Accessed 3 Jan 2017
28. Tyler, M.H.C.: Online Dispute Resolution. in E-Justice: Using Information Communication Technologies in the Court System: Using Information Communication Technologies in the Court System, IGI Global, p. 87 (2008)
29. Benyekhlef, K., Gelinas, F.: Online Dispute Resolution. Lex Electron. **10**(2), 43 (2005)
30. Tyler, M.C.: Online Dispute Resolution. https://ssrn.com/abstract=934947. Accessed 28 Mar 2017
31. Hartl, A.: Online Dispute Resolution—A Promising Alternative for Resolving Disputes in the Future?. In The Law of the Future, vol. 2, Maastricht University, p. 12 (2012)
32. Regulation (EU) n.º 524/2013, of the European Parliament and the Council, of 21 May 2013, on online dispute resolution for consumer disputes

Printed in the United States
By Bookmasters